Biotechnology, Education and Life Politics

What should individuals and society do when genetic screening becomes widely available and with its impact on current and future generations so uncertain? How can our education systems around the world respond to these developments?

Reproductive and genetic technologies (RGTs) are increasingly controversial and political. We are entering an era where we can design future humans, first, by genetic screening of 'undesirable' traits or indeed embryos, but perhaps later by more radical genetic engineering. This has a profound effect on what we see as normal, acceptable and responsible.

This book argues that these urgent and biopolitical issues should be central to how biology is taught as a subject. Debate about life itself has always been at the forefront of connected molecular, genetic and social/personal identity levels, and each of these levels requires processes of communication and debate, what Anthony Giddens called in passing *life politics.* In this book, Pádraig Murphy opens the term up, with examples from field research in schools, student responses to educational films exploring the future of RGTs, and science studies of strategic, mediated biotechnology and the lab practices of genetic screening. Life political debate is thoroughly examined and is identified as a way of connecting mainstream education of biology with future generations.

Biotechnology, Education and Life Politics will appeal to post-graduates and academics involved with science education, science communication, communication studies and the sociology of education.

Pádraig Murphy is Lecturer in Communications and Chair of the MSc in Science Communication programme at Dublin City University, Ireland. His teaching and research interests include science communication, science and technology studies, and emerging technologies, biopolitics and media in education.

Routledge Research in Education

For a complete list of titles in the series, please visit www.routledge.com.

99 Towards Methodologically Inclusive Research Syntheses
Expanding possibilities
Harsh Suri

100 Raising Literacy Achievement in High-Poverty Schools
An evidence-based approach
Eithne Kennedy

101 Learning and Collective Creativity
Activity-theoretical and sociocultural studies
Annalisa Sannino and Viv Ellis

102 Educational Inequalities
Difference and diversity in schools and higher education
Edited by Kalwant Bhopal and Uvanney Maylor

103 Education, Social Background and Cognitive Ability
The decline of the social
Gary N. Marks

104 Education in Computer Generated Environments
Sara de Freitas

105 The Social Construction of Meaning
Reading literature in urban English classrooms
John Yandell

106 Global Perspectives on Spirituality in Education
Edited by Jacqueline Watson, Marian de Souza and Ann Trousdale

107 Neo-liberal Educational Reforms
A critical analysis
Edited by David A. Turner and Hüseyin Yolcu

108 The Politics of Pleasure in Sexuality Education
Pleasure bound
Edited by Louisa Allen, Mary Lou Rasmussen and Kathleen Quinlivan

109 Popular Culture, Pedagogy and Teacher Education
International perspectives
Edited by Phil Benson and Alice Chik

110 Teacher Training and the Education of Black Children
Bringing color into difference
Uvanney Maylor

111 Secrecy and Tradecraft in Educational Administration
The covert side of educational life
Eugenie A. Samier

112 Affirming Language Diversity in Schools and Society
Beyond linguistic apartheid
Edited by Pierre Wilbert Orelus

113 Teacher Leadership: New Conceptions for Autonomous Student Learning in the Age of the Internet
Kokila Roy Katyal and Colin Evers

114 Test Fraud
Statistical detection and methodology
Edited by Neal M. Kingston and Amy K. Clark

115 Literacy, Play and Globalization
Converging imaginaries in children's critical and cultural performances
Carmen Liliana Medina and Karen E. Wohlwend

116 Biotechnology, Education and Life Politics
Debating genetic futures from school to society
Pádraig Murphy

117 Vernaculars in the Classroom
Paradoxes, pedagogy, possibilities
Shondel Nero and Dohra Ahmad

Biotechnology, Education and Life Politics

Debating genetic futures from school to society

Pádraig Murphy

Routledge
Taylor & Francis Group
LONDON AND NEW YORK

First published 2014
by Routledge
2 Park Square, Milton Park, Abingdon, Oxon OX14 4RN

and by Routledge
711 Third Avenue, New York, NY 10017

Routledge is an imprint of the Taylor & Francis Group, an informa business

British Library Cataloguing in Publication Data
A catalogue record for this book is available from the British Library

Library of Congress Cataloging-in-Publication Data
Murphy, Pádraig.
Biotechnology, education and life politics : debating genetic futures from school to society / Pádraig Murphy.
pages cm. – (Routledge research in education)
1. Biotechnology–Study and teaching. 2. Biotechnology–Moral and ethical aspects. 3. Bioethics. 4. Biology–Study and teaching–History–21st century.
I. Title.
TP248.22.M87 2014
174.2–dc23
2013045361

ISBN: 978-0-415-70735-0 (hbk)
ISBN: 978-1-315-88688-6 (ebk)

Typeset in Galliard
by Cenveo Publisher Services

For Fionn, Aisling and Siobhán

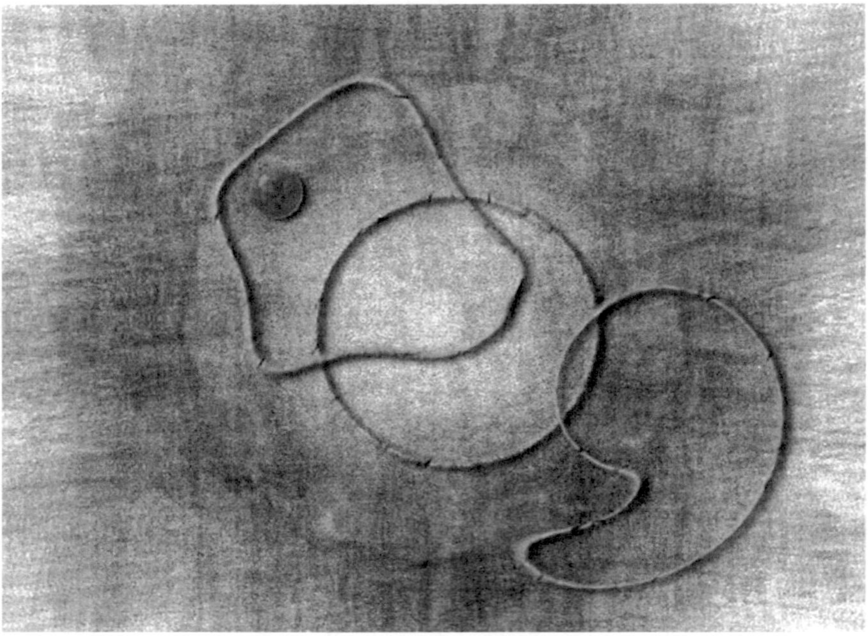

Contents

Tables

Preface

One way or another, this book has been ten years in the making. What started as a PhD project morphed into a mapping out, through case studies, of how future developments in human genetics can be understood and politically argued from the biology classroom all the way to a potential future system of governance.

In Ireland, we have a strong sense of history. It is part of our discourse when things go wrong. We have been inspired by the future, and this too flavours our discourse. Strong histories present strongly unified pasts; much historical non-fiction could only have had one outcome – or at least one outcome is provided from possible alternative histories. Science and Technology Studies often focuses on the histories of the 'losers'.

Decisions have been made, and future generations have a burden to bear. We can say this with some degree of certainty, for global finances, for the environment, and so too, for genetic screening. Perhaps less so for genetic screening of human embryos, and other human genetic technologies, although we do need to tread carefully, even if the practices and discourses of genetic technologies, caught up in market forces, allow us little time and space to pause.

Debates on reproductive and genetic technologies (RGTs) will continue to be divisive, caught up in global 'culture wars' and localised fractures in Ireland, where I write, and other countries once under more influence of the Catholic Church, as well as the US. Here, I map a potential set of pathways into various futures starting with young people's conversations in and around the classroom about reproductive technologies. Future resources will be scarcer, but there will be more choices, a practical example of Strathern's (2006) sophisticated description of the mix of tradition and modernity.

In this book I ask: what new 'good life' of local and global politics can we take from these young adults now that we have on our hands, quite literally, a Brave New World?

Although the school-based part of the field work for this project ended in 2007 for a PhD project – at the end of the Celtic Tiger years – they have codes within their discourse about what we might do next, in the 2020s and the 2030s and beyond. In six schools, we talked about pre-implantation genetic diagnosis (PGD) and therapeutic cloning using embryonic stem cells. There have been many new developments since that in a real way blur the boundaries between technology, culture, nature, human and non-human.

There are three major problems with 'modern biology' in its theoretical implications and its technological applications which all converge in the secondary school.

First, there is widespread distrust of, or aversion to, genetic technologies and research such as genetically modified organisms (GMOs), IVF, prenatal genetic testing/therapy and stem cell research. As these technologies are starting to become a reality in health and industrial fields there are ethical issues associated with their use and many call for tight controls on the basis that there is a line crossed between scientific progress/health benefits and potential affront to human dignity. Others feel strong religious, environmental or cultural objections to these technologies. Some attitudes may be a legacy of the perceived conflict between biological and/or genetic determinism and social/humanistic/spiritual definitions of human nature. Phrases like 'going against nature' or 'playing God' are commonly used. There is widespread opinion among policymakers, educators and the public alike that these issues and possible regulation need to be discussed openly.

Second, there is a decrease generally in numbers of school-leavers taking up the sciences in further education or as a career. Governments and industry are investing heavily in trying to make science more appealing to the general public, and particularly the young. Many initiatives have sought to offer more engagement, sometimes using the hook of new media, 'edutainment' or other methods to inspire scientific enquiry and wonder. The focus for physics, chemistry and engineering-based science initiatives tends to be where things move and explosions happen. For biology, projects aim for forensics and the 'wow' factor. However there has as yet been no comprehensive study to see if these initiatives are working, nor any to see what the young make of them.

A third problem is strongly linked to the second – the biology curriculum in schools. Although changed for the Irish Leaving Certificate for the 2003–2004 cohort to include a strand on 'contemporary issues' which are 'of concern to citizens' (Curriculumonline.ie, 2003) (just 7.5 per cent), it still fails to reflect for students the true nature of bioscience wonder and inquiry for society, or biology's own 'wow' facts, relying on the development of scientific facts and basic school experiments. The relevance of biology to students' everyday lives is not emphasised, nor are the two previous problems outlined above included to a significant degree and depends on the attitude of teachers. This could be called the social and the philosophical of biology. However, many teachers are not comfortable with the use of humanities-based methods of teaching biology, citing it as having no place in the sciences, and many lack training for the type of teaching and assessment more prevalent in the humanities.

Therefore, these three problems with modern biology – a lack of trust in the biosciences, a decrease in interest in bioscience-based careers, and didactic biology education that excludes controversial technological/philosophical issues – provide an ideal opportunity to look at reform in schools. Students aged 16–18 will, for the first time, have to choose biology as a subject. They will soon have to pick careers and think philosophically, even a little, about their lives. They will be future voters, scientists, policymakers, family members who have to learn and offer an opinion on some aspect of biotechnology, which may directly affect their lives.

Students also have other parts to their lives which they may consider more important than biological knowledge, biology education or the career or philosophical implications of biology. They have aspirations, ideas and interests which school biology may or may not change.

The central claim and argument for the book is that biology education should move further to accommodate debate and argumentation about future bioscience and genetic issues, which are as much political as they are scientific or technological. Many scholars have placed this 'futurescaping' into the area of biopolitics, but it is the term 'life politics', devised by UK sociologist Anthony Giddens, that I look more closely at here, to describe how the personal/individual of, for example, genetic screening technologies is also a global future political issue about human characteristics and enhancement, the hegemony of genetic knowledge. New biotechnology needs to include this life politics in its teaching; argumentation, evidence-based opinion forming and debate among students are transferable skills that speak directly to this new terrain of politics, always mixed with science.

Acknowledgements

My gratitude extends back a decade to many people from the origin of a doctoral idea to the development of a book in a similar space. In no particular order, the people who have made this possible are: Pat Brereton, Brian Trench, Miriam Judge, Siobhan Byrne, Barbara O'Connor, Michael Reiss, Pat O'Mahony, Norah Campbell, Brendan Ball, Nicola Baldwin, Eve Merton, Nigel Townsend, Jane Madeley, Danielle Nicholson, Ruth Jarman, David Jackson, Bonnie Green, Ekaterina Kozina and Clare Ashworth. A special thank you to six un-named schools and their teachers. Finally, so many researchers in Ireland – caught up in hi-tech practices and discourses as we are – owe a debt to the philanthropist Chuck Feeney whose funding was a catalyst for educational and scientific infrastructural support at the turn of the century.

Frontispiece: *Society*, an e-painting by Dr. Sabin-Cornelieu Buraga (2013)

Abbreviations and glossary

ARTs: Assisted Reproductive Technologies.

Biosociality: a theory developed by anthropologist Paul Rabinow (1996) describing communities and social groups that are formed around a specific biological issue such as threatening genetic markers.

CAHR: Commission for Assisted Human Reproduction.

ELSI: Ethical, Legal and Social Issues, a small aspect of Human Genome Project but considered a key element in the public communication on the project.

Emancipatory politics: Anthony Giddens' term for counter-cultural revolutions of the 1960s, to free humanity from shackles imposed by gender, race, sexuality.

Epigenetics: the study of changes in cellular phenotype that are not due to underlying DNA sequences.

Expert systems: Giddens' term for those technical systems in contemporary we rely on for authority, to carry out functions and to legitimate, and in which we place great trust.

Frames, framing: method of analysing 'spin', emphasises or omissions in media, particularly news media.

FRDA: Friedreich's ataxia.

HESC: Human embryonic stem cells.

Life politics: Anthony Giddens' term for a politics of choices in 'late modernity', that includes consideration for processes of biological life as well as life-planning and self-actualisation.

Mediation: any process of co-construction between actors, involving technological methods of communication generally. One actor needs to simplify, shorten, or even shout to attempt to get the message across.

Mode-2 science, post-normal science: where 'normal' 'Mode-1' science involved the lone male working in his lab, the new type has several connections and does not operate in linear fashion from scientist output to society.

Modern, Moderns: Bruno Latour's derogatory term for those that insist on staying with Mode-1.

Nature, *as opposed to Culture, Science, Discourse, Being*: Latour's reading of A. N. Whitehead (1920), what he sees as the historically erroneous 'bifurcation' between a 'Nature' out there and a 'Culture' of humans. The reader will note occasional inappropriate capitalisation for some terms which have greater symbolic meaning, some of which are a type of future projection, like Progress, Modern, Science, Nature, Culture, Politics, Discourse, Being, God.

NBIC, convergence science: Nano-Bio-Info-Cogno technologies.

Parliament of Things: Latour's imaginary governance system of human and non-humans.

PGD: Pre-implantation genetic diagnosis – a screening mechanism to identify specific genes in an embryo, where there is a risk of 'defect'.

RGTs: Reproductive and genetic technologies, comprising Assisted Reproductive Technologies (ART), as well as human and plant genetic technologies

SFI: Science Foundation Ireland.

Sociotechnical: an STS term for the Mode-2 situation of societal and technical/material matters caught up together; Latour's new poetic term for sociotechical phenomena is *composition*.

STEM: Science, technology, engineering and mathematics.

STS: Science and Technology Studies (in sociology and philosophy) or Science, Technology and Society (education).

Structuration: Giddens' sociological term for the mid-point and interaction between structure and agency.

Technoscience: the new emerging technologies driving science, unlike the Mode-1 version in which it was the inverse.

Things, *dings*: Latour uses this to describe a political object which requires work from many in a network to come into being.

Witness, witnessing: an STS term referring to a lineage that stretches back to Robert Boyle; a scientist 'sees' a phenomenon (this can be done through media), and creates the conditions to spread the word faithfully and maybe even replicate in laboratory conditions.

Chapter 1

Introduction

Biopolitical control and the possibilities for life political education

Our lives today are controlled as much by the discourses of human biology as they are by biology itself. 'Biology' sounds as controlling as it sounds educational but learning science is not a passive exercise for our young anymore. Nor is it for adults. Science is a critical business in every sense of these words. The sciences and technologies that relate to understanding, creating and developing human life are particularly critical. We are beginning to have the power to alter the genetic make-up of the next generation of humans without traditional methods of reproduction, through decision-making at the embryo stages or direct engineering of the genes to prevent disease in the next generation. Or perhaps for other reasons. There now exists a mix of biological knowledge domains and the social application of this knowledge through modern biotechnology. It is a mixture of both science and discourse. We learn it, we learn from it, we use it, and experience it and respond to it – as animals of course – and much of what we know will change rapidly. I will get back to forms of discourse in contexts most useful to this book later, but let us remain with one common idea of discourse: discussion. The discussion drives the act, which in turn drives the discussion. They become inseparable. Fundamentally, we are changed by, and can change, our biology. This may be the real beginnings of the earliest time we humans can change our future biology, and it may seem like a frightening future. In other contexts besides biological knowledge, we may say that we are burdening our children with a bad legacy of financial debt or with environmental disaster. But it is beyond mere legacy – it *is* their future. Likewise, genetic screening or genetic engineering as well as the surrounding regulatory regimes, the ethical statements and debates, *become* our future generations.

Can we select the genetic make-up of our children, to develop further the concept of the so-called 'designer baby', and should we be allowed to were it feasible in a complete sense? Will it be a free or costly choice like many others in the near future? If there is disagreement with these methods on moral or legal grounds, then begins a discussion, the grounds and means for debate. What if one embryo is chosen over others? How does this square with attitudes to, and regulation for, abortion? How will it be understood and governed by our children? Discussion in such a controversial domain inevitably leads to political debate, in a real sense, perhaps leading to regulatory change. Inevitably, there are strong feelings of politics and passion around germline genetics and embryos (Fukuyama, 2002; Habermas, 2003; McDonnell and Allison, 2006; Stock, 2002). It will continue to be a hot topic in Ireland way beyond the time of writing. This book will explore its global significance by focusing on its implications for future education.

After debate and demonstration, we can then move forward. 'Demonstration', is what Science and Technology Studies (STS) scholars Andrew Barry and Michel Callon call the re-presenting and resisting, the political clash, action in need of an audience, requiring media (Callon, 2004, citing Barry, 2001). The biosciences commonly use systems of internal scientific discussion or external public debates and debate and argument is central to good science, to good teaching and to good politics. Finding out past genetic histories and future probabilities of disease and life chances, ultimately discussion and debate caught in spheres and sites of thought help create a future. Are we ready for a future that brings about significant changes to the human species, or with new traits (Fukuyama, 2002; Stock, 2002)? In the summer of 2012, peak UK viewership for Channel 4 coverage of the Paralympics rose to 5.7 million people (Broadcasters' Audience Research Board, 2012). They watched London on their TV screens, many in amazement, perhaps for the first time realising the wonder of paralympian endeavour. Young people were hopefully inspired by these triumphs over adversity. But are processes of genetic selection of embryos a means of removing this adversity? To be a paralympian means years of training and sacrifice. If embryo selection were prevalent for a range of perceived disorders, would we no longer have 'natural' paralympians? What normative and lab practice decision-making is crucial to the separation of embryos into the 'future potential sick' and the 'future potential well'? So an affirmative biology is, in a way, created around these types of questions and robust propositions, as they occur in books, at the lab bench sometimes, or perhaps even through the data from a computer model; but most likely in the vocalised debates that occur in the TV or radio studio, the courtroom, in parliament, in the boardroom, the internet chatroom, the lecture hall or classroom. This, shall we call it, *proactive* biology may lead us to acceptance, perhaps rejection, of radical technologies that change our lifestyles, practices, our sense of morality, our relationships with an ideal of Nature or Politics, our being, who we are, and who we may be, and to whom or what we connect. But we debate it all too seldom. We are part of biology-as-discourse from the beginning for our citizens in the secondary or high school classroom, and how we might see, debate and shape the future (so often in these contexts, unfortunately, the Great Western 'We').

Central to these debates is a public type of science communication, a *public pedagogical* communication. *Pedagogy,* as used in this book, is quite an academic term. I use it instead of the less nuanced word 'teaching' to exposit how the study of teaching and learning might respond to this biological-discourse control. Also used is the specific example of teaching through social and decision-making arguments about the many reproductive and genetic technologies (RGTs) that are emerging, like genetic screening, therapeutic and reproductive cloning and human stem cell research. There is a much broader concept of teaching at work here. We need our next generations to know, and be involved with, these debates. Science education, and the science teacher specifically, has rarely had the opportunity – nor indeed the professional incentive – to pause and give serious thought about the profundity of bioethics, the sequestered taboo of what has come to be known as *biopolitics*, controlling body and life processes, the strategic funding and political mechanism of genetics and genomics, or the possibility that genetic screening may 'always already' (Fuller, 2011) be veering toward eugenics. Indeed how could, or should, teaching staff ever need to know about the new outlandish landscapes of the posthumanities (Agamben, 1998, 2000; Campbell, 2011; Esposito, 2008)? This is different than modernity, the project of progress. Is this all too alien for the science classroom?

It is proposed in this book that education of young people is the starting point of this new public biology pedagogy, by learning science through argument and opinion-development, the tools of rhetoric and politics. *Life politics* is the concept I have chosen to demonstrate this, as put forward by the sociologist Anthony Giddens (1991). Life politics is not just political action around life processes such as embryo technologies, gene selection or genetic engineering. The description 'life' makes a positive statement, whereas 'bio' is a clinical scientific signifier that can also become a prefix of fear, with the exception of words like bio-degradable and biofuel, many others carry the risk factor – 'biohazard', 'bio-terrorism', even 'biotechnology' itself. *Bio*politics has a history in theory suggesting control, while life politics is a fight back from below. Life politics also consists of life-planning and future living (ibid.), habitats, societies and cities, how personal style and public politics are intertwined (ibid.). Life politics in this context can challenge a social engineering that might misuse genetics in society, or enforce forms of eugenics, the supposed progressive assistance to human evolution developed by early biological engineers of society (that is, social order), William Paley and Thomas Malthus (Fuller, 2011). Paley was seen as an early proponent of universal design in theology. Malthus brought the world to account on its expectation of unlimited resources for non-interrupted global population growth. This is a life politics beyond the life sciences. It is not enough to learn content, to practise science from the school lab. We will get glimpses in this book of how inadequate this is when faced with the real politics and crisis of existential shock and the need to act when faced with serious illness or affliction in the next generations. This is a teaching, learning and public pedagogy that moves these associated issues of human design, finitude, birth, disability and abortion into society, away from 'bioethics' and into education as politics, as argument, as transferrable communication skills. Crudely described, it is TEDx biology.

This combined science and politics – for all that is presented to us in science and media, from pharmaceutical industry content, from business drivers or from misconduct cases or self-diagnostic manuals, or even CERN discoveries and novel crops – has science not been presented within society as 'fact', as distinct from 'mere opinion'? Not if we look more deeply at these issues from the previous sentence again – the biosciences have already been immersed in opinion and politics. The fields of Science and Technology Studies, and new thinking in science communication based on solid research, show us that the context, emotion, affectivity, rhetoric and style *re-presented* are just as important to the teaching and understanding of science as the content (see Ziman (1996), Merton (1973), all the way through to Latour (1993), Haraway (1997), Ravetz (2005, 2006) and Fuller (2010, 2011)). This is a direct result of the messy, social nature of current science and technology. Argumentation, reasoning, persuading, but also controversy. Historically, the discursive side of argumentation for bioscience knowledge has immense importance. It is *loquutio,* speech that determines 'what is useful and what is harmful', separating the 'just' from the 'unjust' which has occurred from Aristotle to Aquinas or the mediaeval *scientia* which were persuasive forms of knowledge presented through thought experiments (Fuller and Collier, 2008). Politics demands a series of solutions to problems; argument is just the beginning of this process. Giddens' life politics tells us that technical processes since the Enlightenment have forced the removal of the taboo from our daily discourse, the unclinical, the messy. However the messiness forces its way back in. Life politics forces us to choose, because it is the politics of choices which easily applies to issues of life itself such as end-of-life, beginnings of life issues and the ethics of life processes. Through these the taboo and the socially hidden re-emerge. It takes *realpolitik* and a critical pedagogy to deal with the mess (Giddens, 1991).

The TED mode of pedagogy is with us, but there is more disruption, more need of debate. We are now in the era of what has become known as germline technologies or inheritable genetic modification (IGM) – changes in genes of offspring that will carry on into future generations. The term reproductive and genetic technologies (RGTs) is used quite frequently in this book, a collective term borrowed partly from anthropologist Jeannette Edwards (2002). These include IVF, genetic screening, therapeutic cloning of embryos, but could also include other forms of genetic enhancement and genetically-modified organisms (GMOs). The main focus here are those methods by which a sufferer or carrier of a genetic disease may eradicate it completely from future offspring. This is the domain of pre-implantation screening (PGS) – where searches are made for broad genetic deletions and anomalies which increase the risk of miscarriage – and pre-implantation diagnosis (PGD) – targeted screening to find a specific set of genes. Biopsies are carried out in fertility clinics and the embryo or blastocyst (Chang *et al.,* 2011). However, any genetic change in an individual embryo for the future person also impacts on future generations. Testing occurs in molecular genetic material, so the cystoplasm is involved as well as the nucleus. In fact, in terms of where research is going, the gene is not even where it is at anymore. Much of the technology has moved from DNA in the nucleus out to the cell itself, in areas known as epigenetics[1] and in proteonomics where proteins' role in disease are studied. Molecular genetic testing is commonly requested for diseases such as cystic fibrosis, Duchenne Muscular Dystrophy, Huntingdon Disease, and Fragile X Syndrome (Barton, 2012). A less common disease, Friedreich's ataxia, or FRDA, is the main genetic-linked disease covered in this book.

This is no longer a time of In Vitro Fertilisation (IVF) and assisted reproductive technology (ART) techniques; that is, of technological contraception, helping women to conceive through technology. New contraceptive controls emerging in the 1960s were so much a liberatory force for change for the feminist movement, in Ireland as elsewhere, yet also paradoxically, feminist theories warn us, this was the beginning of a structural technoscientific invasiveness of the woman's body (Edwards, 1999a, 1999b, 2002; Franklin, 1999, 2003, 2006; Strathern, 1996) (also see Haraway (1991) and Harding (1986) for broader feminists approach to technoscience). We have gone much further. IVF was a first step, now a relatively routine procedure, for clinicians and doctors, if not for couples or individuals seeking this service (Inhorn and Birenbaum-Carmeli, 2008; Slowther, 2008). Other techniques may also soon become part of routine choice for society – gene therapy, which is a germline modification (effecting the gamete, that is the sperm or egg) or somatic (all other cells of the body). Latest developments have also included human embryonic stem (ES) cell for research and potential treatment. Feminist critiques of these technologies contest the patriarchy of the ideals of kinship (Edwards, 1999a, 1999b, 2002; Franklin, 1999; Haraway, 1997), but also the issues of an all-controlling technology, what De Certeau (1984) might call technology as strategy, the totalising power of 'the Proper'.

Cloning, for example, is a disruptive idea in popular culture and society. Parthenogenesis is the scientific term but the popular culture and literary idea for humans is the double, twin, copy, the dőppelganger or a too-familiar other, or closer than that, a non-I or a second self. In biotechnology practice, cloning is a common process of replicating a cell. Under laboratory conditions, fragments of DNA, promoters or molecules can be replicated and done so quite radically using a process called polymerase chain reaction (PCR) used in forensic investigations where there are DNA traces. The media coverage often centres on a process called somatic-cell nuclear transfer (SCNT) (Kfoury, 2007), but more commonly known as cloning, either human reproductive or therapeutic (ibid.; Tannert, 2006). In therapeutic cloning, nuclear material is transferred from a somatic cell to an oocyte (egg) where the nucleus is removed with the aim of multiplying up cells with the same genome as the nuclear donor (Kfoury, 2007; Hayry, 2003). Reproductive cloning of humans, for issues of safety as well as ethics, is universally banned. Some cloning of animals has taken place, most famously Dolly the sheep in 1996. Therapeutic cloning has been considered for the treatment of Alzheimer's disease, Parkinson's disease and Duchennes muscular dystrophy. Embryonic cells are quite potent, or pluripotent, essentially more malleable and programmable, with the potential to be converted into a range of important body (somatic, or non-sexual) cells. Genetic screening by means of pre-implantation genetic diagnosis (PGD) is a technique where there is screening of embryos for a prospective parent, offering the patient the opportunity to make reproductive choices – choosing which embryos or whether or not to allow a particular embryo to go full term. Both of these latter RGTs will be a particular focus here, as the concepts were explored as a topic of discussion with young people, as well as their scoping of future humanity, compositions (Latour, 2010), their mangles of practice (Pickering, 1993) through education and politics covered throughout this book.

There is a life politics around us and it is not just a political science of life. When Anthony Giddens defined life politics, he was describing life-planning, where personal life and public play become blurred, an extension of twentieth-century emancipatory struggles of race, gender, sexuality, environment and worker's rights into embodied self-actualisation and a new supposed autonomy and authenticity for the Modern Self (Giddens, 1991). Although not well theorized in *Modernity and Self-Identity*, he meant, to a lesser degree, RGTs (ibid. 217–23). What could be more political and less coldly bio-educational than changing the human species?

And then there is Latour's odd notion of a once and future parliament, a Parliament of Things, made up of our propositions and vision 'things' and objects (a very specific reading of a 'thing', a partial object read and enacted in multiple ways) (Latour, 1993, 2004a). Odd though it may be, we will be giving this notion particular attention in this book, a politics beneath the national politics we tend to see represented in media.

Besides the future generational impact, the other contentious issue is that an embryo with the genetic markers for such diseases is not allowed to go to full term – it is either stored or destroyed. Here is the crux of genetic screening ethics for many. The human embryo has been a particularly contentious subject in these discourses worldwide because of religious edicts that state it is human life. Most technologies involving the embryo will destroy it. Because of the unique – and controversial – status of the embryo in Ireland and elsewhere in Europe, this book focuses mainly on those types of RGTs involving the embryo. There are implicit or explicit moral boundaries constructed by, and for, governments, policymakers and students (and at times teachers) to describe this entity, the social and philosophical implications of wider genetic research, and most pertinently for local educational practices in schools, what elements might be explored in senior level science education. It is difficult to exaggerate the clash of worldviews on the status of the embryo; it is a symptom of dead-end or circular debates. It is enforced certainty, the inability of experts on all sides (theologians, ethicists, clinicians) to deal with uncertainty (Polkinghorne, 2004). Typically for any sociotechnical issue however – bioethical or just ethical with biosciences attached, if you will – where there is a clash between an instrumental, science-only approach to a diagnosis, treatment or application and compositions of value systems, the repressed issues spill back in – to quote Italo Calvino the 'more enlightened our houses are, the more their walls ooze ghosts' (Appleyard, 2004: vii, quoting Calvino, 1987).

There are other cultural flashpoints around biotechnology. If we were looking for common patterns of unease with these technologies, we would not have to look far. There are concerns about the Nature/Culture divide, and indeed humanity's place in the world. When the European Commission began discussing genetically-modified foods following field trials, anxieties ran deep, particularly in Europe and the US (Bauer and Gaskell, 2002a, 2002b; Hornig Priest, 2001; Jasanoff, 2005; Nelkin and Lindee, 2004). Biotechnology, as the controlled manipulation of life processes, began as a serious option in the food industry in the 1970s, particularly in the forms of genetic engineering and recombinant DNA techniques as they were then known. However, once the widespread release of GMOs was discussed, the debate began in earnest in Europe. There are also clear concerns from a wider legacy of distrust of science or the science industry itself and its collective intentions, for example the issue of patenting (Barr and Thompson, 2002; Hipkins, 2002; Kolstø, 2001).

Jon Turney (1998) has invoked the Frankenstein myth in popular consciousness and indeed Shelly's classic has such cultural resonance running through the ages, the myth of 'Man', monster and Nature. Genetic modification and reproductive technologies such as human embryo stem cell research (HESC), cloning of embryos, and pre-implantation genetic diagnosis (PGD) are on the one hand central to the public image of scientific progress but they also invoke images of humanity 'playing God' (O'Mahony and Schäfer, 2005), a motif that emerges in discussions with young students in Chapter 7. There is no doubt but we can say that these new identities are created from ideas and *things*, these new *compositions* (these latter terms both in the Bruno Latour sense, as we shall see), from clones, chimeras (animal-human admixed embryos), designer babies, saviour siblings, spare parts. They represent a threat to the person on many fronts: interfering with nature or life, or threatening human dignity, rights, religious beliefs, human identity, and health as well as the environment. The very concept of identity, and indeed humanity, may already be changed – does this mean the core of human values, rights and dignity will also (Fukuyuma, 2002)? Issues of identity are most certainly raised by future thinking developments in genetic screening, IVF, and gene therapy. They are areas of serious concern for many globally.

Leon Kass, President Clinton's bioethics advisor, apparently introduced to the fields of social study the technical term 'yuck factor'. This yuck factor seems to mean the sometimes inarticulate expressions and types of revulsion against the norm. It is an adequate expression for the unclean monsters that are associated with the human body and its potential distortions, the Golem/Frankenstein allusions. The issue with this of course is, that in American politics as in other states around the world that call themselves liberal democracies, 'yuck' might summon a constructed public revulsion against many things, including gay marriage, rap music and sex education. So when what are perceived as historical and cultural 'drags' on progress appear in the form of an existential body shock – the creation of a monster, the disruption of the body, the ending of life – politics kicks into another gear.

Biotechnology has also been analysed as an organising force for nation-state building, especially in Europe, involving economy, academic/industrial knowledge, concepts of life and strategies of the boardroom and parliamentary politics (Bauer and Gaskell, 2002a, 2002b; Jasanoff, 2005). We look at this more closely in Chapter 3.

Science is this great Western journey to Progress. However science, as popularly described and understood, is replete with public risk. If Copernicus moved humanity from the centre of the universe, Galileo made us witness the imperfect heavens through the telescope, while Darwin cast us humans adrift into the outer regions. Dawkins probably has done even more to relocate the human. But it is to Martin Heidegger (via Kant) we turn for that seminal move in philosophy into the extreme technological, and for many theorists, toward the posthuman. While Heidegger has been the foundation on which much great work has occurred on how humans as species and Being can co-exist, for Campbell and others, his work has led directly to a move into a darker politics. It is not life politics as we describe it here, but its opposite, *thanatopolitics* – death politics rather than *bios,* the latter of which is the striving in politics and biology for what is good. Technology is the 'tear in Being' (Campbell, 2011: 7) started with *Parmenides* (Heidegger, 1992[1942]), the act of the unfamiliar, a cold ordering while simultaneously, and in contradistinction, orchestrating a move toward nearness[2], comfort of the familiar, motherland even, opening a whole area where 'we' need to proceed with care in terms of state and control, the giant body-politic Leviathan of Hobbes (Shapin and Schaffer, 1985). Campbell's reading of Agamben sees Heidegger's 'saving power' as that which is a power *over* life, while the biopolitics that writers like Esposito advocates, according to Campbell, is the power *of* life (Campbell, 2011: 17 citing Agamben, 1998, 2000). There is a clear tension between a negative biopolitics, in terms of power of control of life and death of citizens, and the biopolitics of life flourishing, which I call life politics. RGTs fit into this political philosophical tension.

The field of STS introduces the relationship between non-humans, humans, objects and spaces. Latour is the main protagonist keeping us displaced in this world of networks and compositions containing quasi-objects, 'things' formed from objects, bodies, discourses, aesthetics, politics, propositions and ideas. The (pre-Republican state) Irish scientist Boyle's contribution to this new idea of the real of everyday objects and things and their effect on our understandings of nature was a fascinating contrast to political scientist Thomas Hobbes, described wonderfully in the game-changing book *Leviathan and the Air-Pump* (Shapin and Schaffer, 1985).

We are now at the theoretical crux of our thesis here – what we need for science communication and education of the biosciences is the argumentation of ideas within the continuing sociology of biology. Life politics is a return of the repressed (Benton, 1991), or how we educators and students can do biology responsibly, where 'responsibility is responding to a call' as Latour said, speaking at a Celsius symposium on compositions and critique at Dublin City University (Latour, 2012; also see ibid. 2010). The shadows hanging over this whole area include the physical bloodshed in the name of race and blood itself. In a fascinating book, geneticist Steve Jones takes us through the history of human progress through religion, royalty and above all, the urge of bloodline which excludes so many, the 'good', the 'bad' and the 'blue', an ancient way of biomapping pasts and futures of kings, species, race and families (Jones, 1996; for STS reading of this topic, see also Rose, 1997). Contemporary posthumanist theory maps this further, drilling down into the molecular level, where exclusionary processes of states and collectives continue with the age of biotechnology, in the immunity or *immunitas* of Esposito, (2002, 2008) and the *bare life* of Agamben (1998), a new prism of inquiry necessary since Foucault's analysis (see also Rose, 2007). It is as if our very atoms can now be isolated and privileged or ostracised for the greater good of society.

There are ambiguities and conflicts in legal positions regarding the strengths of genetic ties (biological parents) or familial ties and kinship (Inhorn and Birenbaum-Carmeli, 2008; Edwards, 1999a, 1999b, 2002; Franklin, 1999, 2003, 2006; Strathern, 1996). While surrogacy and gamete donation is increasing, cases like the recent Irish court ruling where biological parents are granted privilege in an adoption case hark back to a more traditional, bloodline trajectory. In an opposite event, a Florida child 'divorces' his biological parents in favour of foster parents (Strathern, 1996). What Strathern takes directly from such legislative tangles is this – the law must recognise the paradox of biological 'certainties' *and* the social negotiation of new family constituents (ibid. 42 citing Dolgin, 1990: 4). According to Fuller (2011) the nineteenth-century judiciaries, legislatures and executives across Europe each would have brought this thinking towards Divine Providence. We must also strongly consider the site of Hobbes' Leviathan, a living organism of state, science and order (Shapin and Schaffer, 1985). Fuller's aim is to map out comprehensively – and he does so exhaustively – the converging object of social policy and science (helped in no short measure by a structure of social science, social policy and religion) all leading directly to a postmetaphysical and physical development of a Humanity 2.0 (Fuller, 2011). Human enhancement and future families may have new innovations of body and gametes, but old ideas of blood will remain. Old and new ideas of the things we know and see mix together.

And what then of how young people, future generations of prospective parents, see these technologies as they become more visible? Or even those at third level? It has been possible – but should be no longer – for a biotechnology student to go through school and college without ever having heard of the debates, fears and social processes I have just outlined. I was that blind biotech student myself in the past. State, Church, Catholicism countries – all have had their own impact, their fears of biopolitical issues in education across the world. This has caused educators and education researchers to reappraise the way that biology is being taught in schools (Kolstø, 2001; Levinson, 2010; Michael *et al.*, 1997; Parker, 1998; Simmonneaux, 2000, 2001; Reiss and Straughan, 1996). Simple identity and relationships may start with the obvious: 'what is living?' and 'what is non-living?' The traditional text-book answer was relatively clear – the new answer should leave more questions, and we will look at some of these mid-ways through Chapter 3.

Healing or enhancing with reproductive and genetic technologies (RGTs)?

Healing or enhancement: is there a difference?

It is the question that will keep emerging. As much research has shown, particularly in anthropology and STS, RGTs have broken down many barriers in how we perceive human/non-human, living/non-living, nature/culture, as well as previous assumptions about parenthood, family, gender, and the body itself. For policymakers, clinicians and doctors there is a clear question with no clear answer, despite attempts to construct one: where is the dividing line between socially acceptable or moral intervention of genes for the common good and a socially unacceptable or immoral application that goes against some shared sense of values? What is 'the good life' if we cross some Rubicon of changing human biologies? Many moral philosophers have consulted Kant and others to find absolutist values beyond religion or politics. But we are dealing with global cultures here, not to say new forms of life in the philosophical sense and possibly in the literal sense (Rose, 2007). Habermas (2003) uses a metaphor when discussing genetic engineering and the creation of genetically engineered embryos, borrowed from other such philosophers: what is the difference between 'the Grown and the Made'? There are age-old assumptions between nature and human stewardship in that question which proponents may need to discard.

The Good Life or the 'Big Genethic Question' of therapy versus enhancement is actually a whole range of questions dealt with in classroom discussions of some years ago, reported later in this book. Among them are: genetic essentialism; stigmatism and exclusionism associated with genetic predispositions; access to one's own genetic information as healthcare 'consumers'; patient autonomy; current generation/individual human identity/enhancement; the philosophical and ethical implications of perceived 'tampering' with a sense of humanity; human dignity; the religious dimension, interfering with God's or Nature's plan; the mediation and expression of these concepts through film, literature, news coverage, blogs; the political representation of these concepts, and our knowledge and positioning in 'opinion'; future generations as current political capital (and future subjectivities as part of a 'political assembly' for such matters); the autonomy of age and gender, for example testing of children under sixteen for disease markers.

The recent philosophies connecting with these issues are extensions of Foucault's biopower and biopolitics. I am thinking here of Georgio Agamben and Antontio Negri, but also those we give considerable space to in this book for their presentation of a politics of life overcoming its opposite – sociologists Nik Rose, Steve and Hilary Rose (the latter two are married to each other, but no relation to Nik!) (Rose and Rose, 2001), anthropologist Paul Rabinow and the emergence of counterpoints in Italy to Agamben and Negri, Roberto Esposito and Timothy Campbell. There is a well-worn path in the social sciences on biopower and biopolitics (Rabinow and Rose, 2006). By biopower, we tend to mean the powers of biopolitics post-nineteenth century, a governance evolved over time after feudal societies, which became a new power over life. The two poles of biopower, according to Foucault (1978) in *The History of Sexuality*, and giving more detail in his Collége de France lectures, are:

1 anatomo-politics – maximisation of use of bodies, integration to systems, through prisons, hospitals, schools, the way we act in socialised, embodied ways. This he called *discipline*.
2 the species level dealing with populations, births, deaths, longevity, mortality, morbidity. This he called *regulation*.

Both of these poles present a technology in its broadest sense, inclined 'to invest life through and through' (Rose, 2007: 53 quoting Foucault, 1995[1977]: 139). For Foucault, neither, paradoxically, are biopolitics of sovereignty, nor is state an entity in how we have come to know the concept historically, and we will look at these connections in Chapter 2. Bodies get pulled away from the sovereign. The two poles above get closer in late modernity with individual hygiene, health plans, and habits and behaviours influenced from above in urban spaces. The ability, as opposed to 'power' in the sovereign sense of the individual to negotiate new subjectivities that change identities and bodies within these spaces are already controlled to a degree by biopolitics. What the subject creates is 'a vital politics', a life politics. Historically, the strategic biopolitics *in extremis* became eugenics. Habermas (2003) sees a positive eugenics in what is unfolding in RGTs, where individuals are lead by market forces. The increased privatisation of genetic screening services is a threat to humanity as individuals and species. (Thacker (2005) asks that we consider Marx's species production here, a political economy approach.)

While Buchanan *et al.* (2000) ponder on what we do about our decisions as we move with technology 'from chance to choice', Dorkin suggests: 'the line between chance and choice...is the basis of our value system' (Habermas, 2003: 24 quoting Dorkin, 2000: 39). For Habermas, the shifting of the line between both creates two levels of existential and moral reasoning on the subject of PGD: 'self-understanding as moral beings' and 'anthropological background of an ethics of a species' (Habermas, 2003: 28–9); so somewhere between autonomy of the individual (genetic self-transformation) and 'equal respect' for members of a species. We may be 'responsible authors of our own life history' (whether we know it or not) (ibid. 29).

There appears to be a common argument here for an ethics of both species and individual. It has echoes of Foucault. As we continue our investigation here, perhaps morality is a more appropriate term rather than ethics, moving up towards structure, as individuals have little time to pause and reflect. Parents facing a reproductive decision may have to act fast. Nik Rose tells us that there are 'technologies of optimisation' ready to help, quoting Ian Wilmut the lead scientist who cloned Dolly, that we have entered the age of 'biological control' (Rose, 2007: 16 citing Franklin 2003: 53). In wider STS, Foucault's biopolitics is newly interpreted by examining the molecular and somatic agents of biomaterials, life material (Rose, 2007), the movement of, and power over, 'life itself', the Foucauldian term from anthropologist Sarah Franklin (2006).

Healing or enhancement: how do we teach about them?

Why raise all these issues in school? Put simply, as we have seen, these are questions of control with consequences so profound that they need to be addressed as a set of critical educational questions, such as: how much control can individuals have over the genetic make-up of their children? How much control is bequeathed to future generations, and how much to current and future regulatory systems of governance, public and private?

The applications of biology and social knowledge have become separated. Ernst Mayr called it the Weissman Barrier (Fuller, 2011). The consequence of this split is that biosociality and biopolitical debate cannot exist when biology is separated so distinctly from knowledge of society. Although in ways, new forms of life and how they might be governed in various visions of the future are part of current science education programmes, they are just not dealt with in a reflexive manner. For example, in Ireland, Government policy on science depends quite heavily on its available figures for future skills requirements (Expert Group on Future Skills Needs, 2013). To foreshadow for a second what is raised in Chapter 2 on the nature of state-industry-academy nexus of biopolitical control and *bíos* in education, the history of this goes back to Whewell's eighteenth-century university systems in Britain. We will be tracking some vocal influencers of the biosciences in recent history, what I call the 'bio-orators', and how thought collectives developed in the early stages of eighteenth- and nineteenth-century bioscience became the monolithic, millenarian entities at the turn of the present century, where DNA and gene remain icons (Nelkin and Lindee, 2004). We briefly examine countries where science curricula have had religious, nationalist or eugenic principles in their early development, and these ideologies filter through. But in strange ways, countries like Ireland leaped from theocracy to neoliberal knowledge economy quite quickly, with no pause for reflection in between as other European countries, including Britain, have done. Ireland has turned to marketing-style dissemination, 'education and outreach'. Scientific issues of controversy like RGTs have not filtered into curricula as controversies. However, as argued throughout this book, the learning of argumentation on every level about how future humans may exist as part of existing understandings of life-planning and identity is the most crucial element of all. We need to ensure that we are attending to identities in education, constructing and legitimising knowledge, while at the same time teaching to distinguish hard-fought knowledge from internet trolling.

It is to dealing with this at the governance level that we next turn, by way of introduction.

Healing or enhancement: how do we govern with them?

The involvement of state control in against individuals or couples is not a straightforward one and it is the tension that is almost central to any current descriptions of RGTs. State involvement in eugenics programmes is an obvious historical example of social and genetic engineering via selective breeding programmes and selective education pathways. This was indeed a structural, top-down approach to providing what 'was for the best' by a set of social norms. However, is a transcendent, Kantian good not possible, independent of any era? We live in more autonomous times, mostly because the armed forces and boardrooms do most of the governance work of the sovereign now. Neoliberalism has prevailed, and our social contract with science as with other areas of knowledge and production is severed to a degree. How likely was a *social* contract between imaginary silos of science and society anyway? We will see the social in 'bio' as we progress.

Ethics, philosophy and capital 'P' Politics seem to advance less than biosciences, and their use in education and for policy. For instance, how can ethical/regulatory systems of the present be applied to future enhancement, to posthumans? What about *trans*humans, migrating bio-souls into cyberspace and backed up? It starts with the seemingly irresolvable tensions between therapy and enhancement in clinics, and extends out to bio-control as a desirable or possible social order or social good, regulation, policy or parliamentary politics. What compositions of technoscience, objects and contemporary values guide the actions for RGTs and future humans? If we teach about values and perspectives, where are the governance and regulatory structures coming from now and where in the future? Once we start thinking of a bigger picture – even with sci-fi elements – we invoke the concept of design on future populations, and this brings us into the controversial territory of nineteenth-century biosocial engineers Thomas Malthus, William Paley and Francis Galton – and in the case of the latter, eugenics or improvement of the human species. In fact, as we shall explore in Chapter 2, design is a central concept to biological thought. Evolutionary biologists would be annoyed though to read Fuller's account of William Paley influences on Darwinist design principles (Fuller, 2011: 194–5), for this seems to play into the hands of creationism. Paley used the 'watch on the heath' metaphor, implying a complex design requires a designer. Since the public efforts of Darwin's cousin Francis Galton, and Malthus and the population engineers, we humans are the designers. Malthus worried about over-population but accepted the deaths of many through natural or societal causes. Gaia theorist, and ultimate structuralist, James Lovelock indicated something similar more recently. Concern for future populations is warranted, yet established theories are highly contentious in today's politics.

For the philosophers and futurists there is, to borrow an apt phrase from Aldous Huxley's famous book, a Brave New World. Huxley's grandfather Thomas

Henry was another eugenics-orientated engineer. We have many philosophers and sociologists who look ahead. The optimists, and some-time evangelists, include Gregory Stock, Machio Kaku, Steve Pinker, Craig Venter, Ray Kurzweil and James Watson. The pessimists include Francis Fukuyama, Peter Sloterdijk and Juergen Habermas. Fukuyama, in particular, fears a future dystopia in his writing. Nietzsche and Heidegger are involved negatively in this whole enterprise. Both of these philosophers will have a heavy influence in how we think about biology, technology and humanity in this book. Writing just post 9/11, Fukuyama expects 'modernity' to wipe away radicalism. But the repressed surely returns, the radical terrorism as much as the 'yuck factor' that he himself recognises, and technology both leads and provides the tools. Beck's (1992) risk society may provide the context – but there is no modern progress to cling to.

To stay with a global picture, there is structural context for all sciences not just the biosciences. It is what STS scholars call 'post-normal' (Ravetz, 2006) or 'Mode-2 science' (Gibbons *et al.*, 1994; Nowotny *et al.*, 2001) 'Science' or 'technology' are not easily distinguished in traditional ways anymore. It is the *technoscience* of Latour, and Haraway, a prominent feminist STS scholar, a collection of sciences subordinated to broadly social technological needs, or technological impetus, the science of reflexive modernisation in Giddens, Lash and Beck (all 1994), science-as-intuitions, constant self-confrontation through ethics and risk. Anthony Giddens is a crucial figure within debates about globalisation and institutional science. Researching this requires greater trans- and inter-disciplinarity, between innovation policy, social sciences, emerging technology ethics and the humanities. When Donna Haraway speaks of humans as cyborgs, her argument is largely a metaphysical one, with physical manifestations – human dependence on devices and technology were always in evidence, and increasing (Haraway, 1991); for nano-bio-info-cogno (NBIC), or convergence, researchers, a reality is already created, with wearable technologies among the first to make the mainstream. Areas of self-confrontation associated with these mega-science NBIC projects include the field of anticipatory governance, sophisticated to-ing and fro-ing between strategist and publics on 'disruptive' technologies of the future. Since the Human Genome Project, there have been small investments in ELSI concerns (so-called 'ethical, legal and social issues'), perhaps 5 per cent of the total US National Science Foundation budget, for example, but also funded across Europe. Institutions confront themselves in a small way, while ELSI becomes separated from modern, industrial-higher/education/policy discourses of biosciences – the potential for segregation towards, at worst, eugenics, or at best, commodification, but also potential harm to health and environment.

Another related structural context within science communication is the focus on 'dialogic' models where there is more consideration for inputs of the non-experts at the fringes of public engagement with science and technology. This can extend to consensus conferences and areas of debate all the way to the quite democratic idea of co-design, where DIY-bio 'geek' kids invite others into their garden shed for some genetic engineering using open source 'biobricks'. We have truly moved far into the *Science of Life*, as imagined by Julian Huxley, another grandson of Thomas Henry, and brother of Aldous.

Perhaps the first indication that contemporary philosophy was taking RGTs serious was the eruption of a row in 2000 – following the announcement of the mapping of the Human Genome Project – between philosophers Peter Sloterdijk and Juergen Habermas (more serious than most academic spats!) following a controversial paper *Rules for the Human Zoo* delivered by Sloterdijk which was covered extensively in the media.

It is with Giddens' life politics that I will keep coming back to, with this central point – these issues are political in a traditional sense (Politics with a capital 'P') and in a new sense (small 'p') that science studies scholars and sociologists would describe it – decision-making, representation and presented opinion to public and private forums are central to an understanding of what life is and the novelty and existentialism of studying 'life itself' (Franklin, 2006). Rose (2007) comes down on the side of the techno-realists and against ethics in ways – there are intertwining multiple pathways leading to futures, but no paradigm shift into posthumans, 'after nature'. He sees molecularisation, the site of real actions, as 'a style of thought' (ibid. 5) invoking seminal STS scholar Ludwik Fleck. So it is now a 'somatic ethics' (ibid. 8), meaning belonging in non-sex cells, a salvation, new forms of life in a Weberian, Protestant, capitalist way, as well as optimisation, subjectification and more emphasis on expertise and bioeconomics.

This would be such a grounding subject to bring into biology education – vital political species consideration of individuals and the future, while actually creating this future, even by discourse alone. The central point of the book is that the educational arena is the place to start this engagement with life politics. Futures, future humans, future non-humans in Latour's (2004a) new political assembly. Young people can be a part of this now. This goes deeper than science education as a means of addressing science literacy (important but not exclusively), or even exploring the 'ethics' of RGTs.

At the moment, OECD reports, Government, educators, industrialists all offer a critical indictment of the Irish education system, stating a need to overhaul basic science, maths and engineering, currently known as STEM initiatives. However, the last thing we need to do is sacrifice our future engagement with the big issues of the day for increased technical specialisation. At a time of genetic and climate risk discourse, this is about engaging young people with reproductive decisions and making them part of a political future on the subject. The Public Engagement with Science (PEST) movement is a substitution for what was once 'PUS', that is Public Understanding of Science, a school of thought for scientists and science, whereby 'the public' were deemed to have low levels of science literacy, which could be rectified by greater amounts of knowledge (Einsiedel, 2005). Since about the turn of the millennium there is greater import given to the Mode-2 bioscience just described, pulling in philosophical thoughts of a posthuman future; this is serious multi-way communicating. We can no longer just assume or enforce bioscience communication as dissemination. There are obvious societal flashpoints for consideration in education, but emerging technologies are at the crux, be they 'green' or the NBIC convergence sciences.

The other central question for education and society concerns the old-style debate, speech, dialogue, performing/drama; can it, or how does it, counteract, or contribute? Performing science, it is strongly argued here, is a key transferrable skill for young people, performing in this messy new Mode-2 science (Gibbons, 1999; Nowotny *et al.*, 2001), not focusing on the other type of laboratory science also performed, but the political drama that bioscience brings to public discourse and can use. In Chapters 2 and 6, we look briefly at the macro- and performative, mediated politics of Blair, Bush, Venter and Watson as the Human Genome Project becomes a nation-building and continent–building (or bridging) flagship project (Jasanoff, 2005; Torgersen *et al.*, 2002) important to the bio-orators of Chapter 2 to maintain their authority and for bioscience education.

Can the biology curriculum deal with such lofty issues such as biosociality, materiality/object-orientated research, the ethics of human enhancement, dealing with finitude as a form of modernity and thanatopolitics, the 'posthumanities', but also, crucially, argument and positioning for face-to-face debate and media, the performative skills?

In this context, of all these connecting sociological and biological pathways, I want to emphasise the importance throughout this book of the relevant ideas of Giddens' life politics, supplemented by Latour's compositional work and their relationship to education. In this overview, I propose that a form of life politics is allowed to happen at the formative secondary education level. This life politics is a reflexive self-confrontation for biology itself on both distinctions and boundary-blurring of 'human/non-human', the 'natural/unnatural', 'therapy/enhancement', 'fact/value' and the communication of risk in future life-planning. Life politics is a developing politics, not necessarily new, perhaps that which was 'always already' there, softer than the almost inescapable structures of biopolitics. Why it is 'always already' there is that the task of modernisation 'cleans up' technical process, hiding the taboo of sickness, madness, death, finitude. It is a rebellion underneath, like Beck's sub-politics.

Giddens (1991) formally defines life politics as that

> which concerns political issues which flow from processes of self-actualisation in post-traditional contexts, where globalising influences intrude deeply into the reflexive project of the self, and conversely where processes of self-realisation influence global strategies.
>
> (ibid. 214)

and

> life politics does not primarily concern the conditions which liberate us in order to make choices: it is a politics of choice. While emancipatory politics

> is a politics of life chances, life politics is a politics *of* lifestyle [original emphasis].
>
> (ibid. 214)

In summary, the book presents a case for argumentation about RGTs in classrooms as an essential educational and political space that address the life politics of science-and-society primarily coming back to the key two life political themes of *witnessing* and *mediation* (see the end of Chapter 2 for a detailed review). Life politics as I present it here is a linking of current decisions on human genetics to a political future, and back again to the child in the classroom.

So the central point is that the use of RGTs, and contributions to debate on RGTs, are issues that are fundamentally communicative and political. The way we choose the discourse maps out how we choose these futures. Even a rough straw poll in a class from six schools gives us a small impression of what we might crudely call – without quantitative or qualitative verification – 'attitudes' on future use of currently controversial technologies, which may remain as action, were these young students to find themselves in these terrible, or exhilarating, situations. We cannot predict action and trends as futurologists attempt to do. We cannot say what any individual, or couple, will decide to do in future, faced with a terrible decision. We can 'play' out the arguments, while making 'real' decisions. This is a different type of pedagogy, a different way of looking at the future of socioscientific and sociotechnical education for young people.

It is important to point out – and this will become clearer as we move through the chapters – that the classroom discussions conducted as part of doctoral research were done at a time when the objective was to elicit basic responses to Big Ideas of Bioscience outside the curriculum, as it were. The young people were shy, awkward, caught in the glare of a new approach for biology class. Hormones, thoughts on lunch and music were competing with a heavy curriculum of work. Where there are short accounts of the work done for a PhD some years ago, they represent baby steps for an adolescent, away from Sloterdijk's (1988) *kynicism*, humour with laconic non-critique; the postmodern attitude of life political self-actualisation (Giddens, 1991), away from the stasis of practices, but not so close yet that it can be described as elite argumentative discourse, or even formal, or especially 'modern', discourse. Some of these young people would be considered disadvantaged, but not always in the ways teachers originally thought.

RGTs are collectively a type of 'technology' of security (Campbell 2011: 112), like many others that offer a strategy for a state on forms of life, for populations that fight a war on disease for the presently living, like cyber-surveillance and military security, like remediation of climate change, like the welfare state, like the extremes of *aut aut*, deciding who is allowed to live and let die (ibid. 112). These immunitary processes, these *technia*, are existential cloaking and harm-protection against the ever larger and more complex levels of risk data (Beck, 1994; Giddens, 1994; Lash, 1994; Wynne, 1992) as uncomfortably collected by the processes of the *witness* (see Schaffer and Shapin, 1985 and Arendt (2006) [1963] for versions of this prevention of totalizing technological solutions where the life and the human become devalued), both seen and unseen life; and *mediation*, combating through the voice, disallowing a thanatopolitics, or *death* politics, to override that of life.

The workshop sessions

Between 2005–2007, pre-recession, my work brought me into many secondary schools, six of which were picked at senior level, with students between the ages of fifteen and seventeen. This was the Celtic Tiger; these were the Celtic cubs. The piloted work programme I introduced to schools was based on Wellcome Trust resources, the educational film *The Gift* and its complementary role-play/debating activities, which was once only found at the location www.geneticfutures.co.uk. The educational film has now been integrated into a full interactive training resource by Y-Touring at www.geneticfutures.com and www.theatreofdebate.com. The fifty-minute movie was played in its entirety, followed by a short discussion, then the centre piece which was an interactive debate where students were asked to move out of their desks and form an imaginary line in response to statements about character's degree of 'being right' or their motivations. The imaginary line was a five-point Likert scale from 'strongly agree' to 'strongly disagree' across the classroom floor. The students took up their positions and argued their case.

If…Cloning Could Cure Us, a BBC dramadoc, was later added to class discussions and similar statements created based on Wellcome Trust teaching materials developed by Nigel Townsend (1999), and based on a script by Nicola Baldwin (1999). One of two bioscientists from the National Institute for Cellular Biotechnology, Dublin City University also accompanied me on separate days to each of the schools to contribute the 'outreach' bit – here was scientists not connected directly with RGTs but tangentially networked in the institution of modern biosciences. The text refers to them only as Presenter A and Presenter B. The six schools were spread across the province of Leinster, codenamed to preserve confidentiality as North Dublin (ND), South Dublin (SD), North Dublin City (NDC), Louth (LH), South Kildare (SK) and South Wicklow (SW). There were about 200 students in total who contributed to the original work for PhD study, including the pilot elements.

The empirical work of these school practical workshops is complemented, and supplemented, by extra empirical work from: Debating Science Issues, an intra-school debating competition on controversial emerging technologies in society; the Interdisciplinary Science module taught at DCU where I coordinate a genetic screening debate activity arising from research work in schools; and onsite observations and notes on reproductive and screening decision-making practices from the National Centre for Medical Genetics in Crumlin (publicly funded) and the Beacon CARE Fertility Clinic (privately funded), both situated in Dublin.

For the film showings and proceeding debates that took place between 2005–2007, *The Gift* (1999) was a modest, somewhat non-future-proof, telling of a story by Nicola Baldwin (1999) from Y-Touring, funded by the Wellcome Trust and the YMCA, with flashback structure of three generations of the one family who must come to terms and deal with Friedreich's ataxia (see Chapter 4 for further reading of the film and plot, as well as the plot and characters from the BBC TV 'educational' movie *If... Cloning Could Cure Us* (2004a)).

As well as the more direct ethical issue with 'designing' future babies, and cloning of humans, the motif of the *gift* becomes an important one, as we shall see later, a concept going back to Heidegger's *Parmenides*, through the 'posthuman' analysis of Agamben (1998), Esposito (2008) and Campbell (2011) the *munus* part of community, which is the 'buying into' some form of community. This community was once thought of local but is now extended in space and time, as well as the collective of humans and non-humans – whom shall we care for, to whom shall we *gift*? This, as Heidegger began thinking about (see also Mauss, 2002 [1967]), is a form of giving of a gift that risks the surrendering of one's individual identity. So community is a gift that is really gift-exchange. The dilemmas, the global decision-making, then comes into the matrix of decisions currently facing worlds in crisis within a neoliberal frame, the autonomy of the self that Giddens writes about that is pitted against some notion of collective responsibility, for now or later generations, even as it is witnessed now and mediated later.

The structure of the remainder of this book

This first chapter introduced the complexities of RGT and biotech for 'Mode-2' science education, and classroom discourse, as well as outlining the life politics concept with regard to 'witnessing' and 'mediation'. Debating skills and engagement with biopolitics is crucial for biology education. The introduction set out the issue to be addressed at the intersection of biotechnology ethics, political debate and education. We touched on the Foucault/Habermas preoccupation (odd bedfellows!) of species ethics, versus the embodied discipline, what became a live philosophical issue following the HGP, but never an educational one. The simple coding of this problem as presented to prospective parents is – how do we tell our children about the choices for *their* children? How do these relate to the elements of social and future societal control? We introduced the empirical work in classrooms on which this book is based. The central theories that rarely can brush against the school curriculum were introduced: STS and ethics, but made the more vivid in life politics.

Chapter 2 concentrates on the structural of the political dimension. It demonstrates the dialectic relationship between: 1) the social, cultural and political structures of nation-states, business and biosciences, the 'biopolitics'; and 2) the education systems in Ireland, Europe, and US. This is an important comparative and sociological analysis to identify how and why, while the gene became the icon of the Central Dogma in the twentieth century, and onwards to the twenty-first, the taboo, messy nature of real biology-in-society is missing, 'sequestered' from science education as Giddens would call it.

Chapter 3 looks at public opinion in a fresh way for the biosciences, suggesting that, while it is important for opinion to be formed for argumentation, it cannot be informed with biological literacy alone, from the school room all the way to parliament. The concept of opinion is looked at in the context of biotechnology and bioscience, how both historical political philosophy and contemporary STS has looked at biology and society, and how the vexed question of public and political opinion fits into all of this, specifically those ideas that are argued as thoughts about biopolitics throughout society. The chapter looks more in-depth at the ideas of Anthony Giddens and Bruno Latour as they pertain to our intertwined double-helical concepts of *witnessing* and *mediation*.

Chapter 4 is about identities. The chapter zooms into the lives and practices of some of our participants in six schools – students and teachers. It interweaves the narrative of the two films shown in classrooms during research with those of the participants. Our discourse 'sites' are conceptual settings; the 'locales' are the physical ones, where science fact and fiction interweave; the Beacon CARE Fertility reproductive clinic and the speculative world of sci-fi become part of young peoples' representations of RGTs.

Chapter 5 looks at curricular and pedagogical matters in more detail for biology education, using studies and empirical work. The text critically engages with the potential interactions between the 'two STSs' – Science, Technology Studies and science-technology-society in education.

Chapter 6 introduces the concept of framing in understanding media and audience engagement with media and face-to-face events. There are descriptions and quotes from young people 'moving around' physically and conceptually in debates, and how one argumentative person can influence the others, with or without back-up evidence or validated opinion.

Chapter 7 presents nine potential futures for genetic engineering/screening of humans, by way of discursive engagement between students' own accounts and arguments of scholars on science and society. Genes, embryos and future generations are integral to this controversial mix. The important futures that need to be accounted for within our life politics discussion, as primarily 'futurescaped' by young people in classroom debates and interviews but also informed by framing theories of media/culture as well as 'the Latour set' and Giddens, are described in Chapter 7. Nine genetic futures are outlined: Modern Progress, Harmony with Bifurcated Nature, Fortuna/Fate, Parent/Patient Autonomy, Sacred Embryo, Future Child, Future Humanity, Commodification and Pandora's Slippery Runaway.

Finally, in Chapter 8, the elements of each chapter are brought together to ask how theoretically and practically life politics might work for education and a system of governance. The book concludes by demonstrating the logic of (re) introducing politics into biology. The argumentation of young people serves them well to be critical thinkers, in a changing identity-oriented politics of the future. Further, if we reduce 'presentism' (Thompson, 2010) from these arguments in favour of the concept of reciprocity, rather like sustainability education, we will have begun the dialogue between a sequestered Mode-1 *misrepresentation* of biosciences of the present with Mode-2 Parliament of 'Humans and Things' orientation of the future, which is actually a more realistic view of *present* science, and a more engaged education. The chapter concludes with some practical actions for teachers to connect to this parliament.

Notes

1 There is fascinating research on how certain genes can actually change in one generation and be passed on to the next without a change in the underlying DNA structure. This would have been called Lamarckism in another era, which was a rival theory to Darwin's.
2 The proximal or nearness *motif* emerges again with our look at media semi-obsession with the topic in Chapter 6.

Chapter 2

Biopolitics and state in education

On Thursday 12th July 2012 in the Exam Hall of Trinity College Dublin, a well-dressed crowd waited anxiously in anticipation. Looking down from the walls were portraits of figures of historical interest from the Natural Philosophy of Irish Protestant Ascendency figures; such as Bishop George Berkeley and Edmund Burke, Queen Elizabeth I, Dean Swift and William King, the seventeenth-century Anglican Archbishop of Dublin among others. The audience was waiting for the arrival to the podium of J. Craig Venter. He was about to present a twenty-first-century update of the famous 'What Is Life?' lectures, originally delivered in Dublin by Erwin Schrödinger sixty-nine years earlier. The original lectures influenced everyone he knew, he said. This was a 'reboot', as Venter himself called it in the pre-circulated blurb, in the age of synthetic life. As we apparently stand at the threshold of a new paradigm, Venter asked: 'What is Life 2.0?' It was a central talk at the Euroscience Open Forum (ESOF) 2012 in Dublin, the 'Olympics of science'. The Taoiseach (Irish PM), the Chief Scientific Advisor and Helga Nowotny, Director of the elite European Research Council, were in attendance. Tweeters were poised, ready to communicate to the outside world through the hashtags #JCVenter and #WhatIsLife appropriately covering the 'digital age of biology' in Venter's words. In the Exam Hall of Trinity – what an unexciting name for such a fantastically adorned building! – the dusty history of the past meets the futuristic present. Two institutions originally housed Dublin Institute for Advanced Studies (DIAS), Trinity College Dublin and University College Dublin, and the patron of the Institute was the then Irish President Eamonn de Valera. UCD had the reputation in Ireland, long gone, of accepting predominantly students from a Catholic, farming background, while TCD, with its long history, was a bastion of Protestant Ascendency in Ireland. Some might say TCD is a small, remaining legacy of the British Empire. Both are currently courting each other through the Innovation Alliance, a node in the short-lived 'Smart Economy' of Ireland. From the Nature of UCD to the Culture of Trinity – the curious relationship the Irish state has had with science. De Valera asked Schrödinger to be Director of Theoretical Physics and he accepted; his original public lectures were hosted in TCD on 5th, 12th and 19th February 1943 as part of this post.

Before speaking, Schrödinger took the Chancellor's chair in 1943. Venter is seated here tonight, flanked by two Lukes, cancer researcher Luke O'Neill and the person in Schrödinger's old role as Director of the DIAS, Luke Drury, seated on either side of the Provost's chair where King George V and Queen Mary were seated on their visit to Ireland 101 years before. Tonight is the Euroscience Open Forum show, a spectacle, performed, and carefully choreographed with history. Politics pervades ESOF. Drury opens, having cause to mention the Higgs Boson, 'sort of' discovered a week earlier (a great PR coup for this event, an attendee muttered). Drury makes a caustic reference early in his opening remarks about the lack of funding to basic science in Ireland, a whole other political debate that was present in undercurrents throughout the Euroscience Forum that week, where Irish state funding for science, in the midst of recession, appeared to be favouring commercial projects.

Before his speech, Venter interacts with Jim Watson, who is bristling uncomfortably. They have had a troubled relationship, and there still appears some element of tension between them. Venter's speech was an important keynote of ESOF 2012 within a composition of process and events that links laterally to the Human Genome Project and industry and stretching back to Whewell, Paley, Malthus, Darwin and Schrödinger himself, past practices of biology and human social engineering. Paley once said: 'There cannot be design without a designer'. While those major figures of bioscience in society saw humans – or more specifically the state, with advice from themselves – as the new designers replacing God, now here tonight at TCD Exam Hall, the design *motif* was taken up by Venter. The convergence/human enhancement agenda always takes the Godhead mantle, the Designer role. (In German, it is *verdinglichen*, 'to make into a thing' (Jasanoff, 2005: 205).)

Venter and Watson appear to be in on this stately occasion, an event that continues the establishment of Ireland as an entity within a larger Republic of Science. There has always been a strange and troubled relationship between biological understanding of citizens and the concept of the sovereign. This second chapter of the book presents the wider Political structures that appear atop of these networks (politics with a capital 'P'), the cultural context of the intersection between the biosciences, academic/industry marketing and education; what can be loosely defined as the discourses of modern biotechnology relevant to how RGT may be understood and governed; and more, the very real entities that have been created and enhanced since the mapping of the Human Genome Project. While the HGP, in a large-scale, mediated way, announced the return of the taboo of 'ethics' and the 'social' around biotechnology in the form of ELSI, education remains relatively uninfluenced by this good citizen twin.

Also there is the macro-political. Jasanoff (2005) and others (see also Gottweis, 1998) have analysed how national and intra-national power structures of the EU and US emerged in a new century formulation through enacting bioscientific institutional commitments (Jasanoff, 2005). In an oppositional process, bioscience structures and strategies have become framed as genesis projects for nation-states, re-creating them, and in cases such as Ireland hit badly by economic crisis, with plans for future return to growth. This nation-state organising was happening in Europe before the financial crisis, even before the biosciences became the big player that it currently is in the EU. Within these tectonic shifts, new organisations of what and how life is emerges. It is not, however, a simple matter of large-scale cultures and politics of life processes impacting directly on how understandings of life and future generations of genetic traits occur. The big structure of culture is also changed by what comes out of, by mediation, the biotech lab. One shapes the other dialectically. Life politics is the reflexive force here. When there is control of bodies, organs, cells, genes – Foucault's biopower in essence – life politics emerges from the undergrowth as opposition and creative destruction. Biopolitics emerged in the EEC when biotechnology was identified as a great European project, legitimising neo-liberal democratic development through science and technology (Gottweiss, 1998; Jasanoff, 2005). This politics has now impinged on traditional states, fastening them to a global order of biopolitics while paradoxically re-creating independent sovereign structures (Jasanoff, 2005), as evidenced by each country developing a unique national platform around Big Science, specifically Big Pharma or Big Bioscience (ibid). Crucially, these processes would not happen without its shadow, life politics, the mediation and argumentation that occurs at local and intra-social site level, connected to formal, and to a certain extent, informal education through outreach initiatives.

There is a tradition here as Venter speaks in Dublin linking Orators and Defenders for biology. We might recall Darwin's battles with Wallace as competitive biologists – there is social enterprise of biology as much as its content (Driver *et al.*, 1996). There was performance and rhetoric after the DNA discovery by Crick, Watson and Franklin in Cavendish, a Rockefeller-funded project (Rockefeller having its eugenics programmes earlier) and later, Watson versus Venter on the HGP, a social enterprise of competition if ever there was one (ibid.). Canguilhem described a new paradigm of 'communications' and 'codes' (Deleuze and Guattari, 1987; Rose, 2007). In the earliest days after Darwin, Galton politicised survival of the fittest, but the history continues today in the culture of media, in genetic selection but also in stem cell debates, climate change, the spectacle of opposition, of public controversy, as dramas played out in Big and Small Media. They are more than styles of thought. They work, we are part of their composition, constructivist thinking and irreductions back to Pasteur (Latour, 1988) and Priestly's 'de-phlogisticated air' (Fuller, 2011: 190), collisions of ideas whose time had come (and were constructed in this way) but ultimately needing its champions. Most of these individuals professed at lectures like this. It was part of the discursive genre (Fairclough, 2003).

For all that Programme Managers (as Fuller (2011) sees current biotech leaders) drive on the enterprises of molecular biology in real practices (as well as in systems and synthetic biology), there are still surely, in the shadows, in the interstices of practices, those who, as Canguilhem saw in Crick and Watson, reside in the power of communication, discourses and reorganisation of information. Meaning, performativity and rhetoric drove the new sciences of life. The all-important mediators of biology. Does Venter carry on this visionary tradition? Perhaps, in recent history of the biosciences. Does Watson? Maybe Dawkins? But there are not many who can claim to be visionaries of the biosciences. Are the programme managers of biosciences the new fame-seekers? Or the funders, the venture capitalists and philanthropists? Are there others, more substantial, more radical, messing with practices at the fringes? Venter's digital biosciences (reminiscent of Thacker's (2003, 2005) biomedia, molecules enmeshed in digital media) are examples of how 'objects' and 'things' are fetishized in modern biology as elsewhere in technoscience – genes, proteins, molecules, prions, DNA, chromosomes. But things have meaning only in human discourse. And things work. Where are these things? They are the *dings* that Latour (2005b) speaks of, the by-products of discursive practices, incomplete, partial orderings needing perspectives, argument and action from living and non-living, with some consensus and more compromise, based on Norse and Celtic political gatherings.

The subject gets lost, with eugenic and extreme instrumentalist possibilities raised. But it is the rhetorical power of capital 'P' politics, the parliaments and halls of debate which have driven forward the often uncomfortable links between biology, social order and the sovereign. Foucault declared biopolitics as that control of individual bodies and populations that makes a break from the power of the sovereign (Esposito 2008: 32–44), introducing the power vertically upwards and horizontally, more controversially (Latour, 1993), as emancipatory studies tells use that power comes from above. But state has involvement, and I argue here, state education must too.

I want to present here a quite brief historical and sociological account of the biopolitics that shapes, and is shaped by, nineteenth- to twenty-first-century political thought. This will be done through snapshots of speaking and lecturing genres, from early theories of race, breeding and bloodlines, the helping along of evolution for the so-called 'able-bodied', to the 'captive bodies' of institutions, ideas and action (lectures, parliaments, media) that directly impacted on current bioscience and bioscience education, particularly the presence (or lack) of 'humanity' and societal considerations in curricula.

Biopolitical control: the influence of the bio-orators in history

The context and contemporary study of biopolitics is generally understood to emerge principally from Michel Foucault ((Foucault, 1995 [1977]; also Campbell, 2011; Esposito, 2008; Rose, 2007). While outlining the related concept of *biopower* as it emerged in the nineteenth century in the first chapter, we made the distinction between regulation and discipline. Regulatory biopower was a systemic control of populations, through birth rates and registers, deaths, illness charts, while disciplinary biopower, on the other hand, controlled the individual through everyday processes of living in collectives, be it at the desk in school, at the cinema, at hospital or locked away in some other institutions sequestered from view of society, to use the Giddensian concept. Biopower, understood as a power emanating from the state, has now got more complicated, and the loci of power, if any, is harder to find in the broader *biopolitics* where there are different meanings, practices, discursive contributions, pulls and pushes of *life* from molecules to media corporations.

There is an implicit collective identification with genetic essentialism through a political, national, and economic rationale that supports the biosciences. Bucchi (2004) describes how scientific facts are consolidated through public discourse to the dismay of those who claim scientific authority, yet the reification of 'facts' are common to both 'scientists' and 'non-scientists'. Today, biopharmaceuticals and biodiagnostics are big business and biotechnology as a political phenomenon is closely allied with economics in media discourse. Ireland's expenditure as a percentage of GDP on science, technology and innovation has risen even since the recession to just less than 1.8 per cent in 2010, and although similar to the UK is eclipsed by hi-tech spends in Scandinavia (closer to 4 per cent in these countries) (EC Eurostat, 2012). The announcement of the completed human genome in 2000 by President Bill Clinton and Prime Minister Tony Blair exemplifies the prominence of state in bioscience, highlighting as it did biotech's use of an international identity with a progressive cultural image, symbolically announced on the first year of the new millennium. Although the genome had yet to be completed in any practical sense, the global biotech industry ramped up its focus on genetic diagnostic treatments for disease around this 2000 announcement. The event coincided with an increase in media coverage of 'gene for' discoveries relating to distinct traits (Nelkin and Lindee, 2004). A common – and growing – self-referential system of nation-state/commerce[1] may well be part of a scientification of modern social life generally (O'Mahony and Schäfer, 2005), just as S&T is becoming increasingly socialised, according to Nowotny *et al.* (2001) and Giddens (1991); that is the instrumental authority for expert systems is diffusing outwards and *downwards* given the traditional prominence of science (Rose, 2007). Yet for many commentators in the sociology of science – among them Wynne (2005), Jasanoff (2005), Nowotny *et al.* (2001), Hornig Priest (2001), and Nelkin and Lindee (2004) – there is a failure by science-industrial systems to either acknowledge an equally growing critique of the life sciences' complex relationship with mediated global forces and globalisation generally. Giddens (1991) and Beck (1992) also critique late modernity's turn to issues of self-identity, personal risk assessment, and decision-making. Institutions rarely 'fess up' to these pathways of control either.

Politics was blamed as an interference during the 'Bush Science' era (Mooney, 2005), the George W. Bush administration standing accused of manipulating and spinning data on stem cells and climate change. Charges were made that this was a form of Lysenkoism, after Trofim Lysenko, the Soviet agricultural scientist who rejected Mendel's theories in favour of his own hybridisation versions, which

Stalin favoured and enforced as a science. However, biotechnological innovations – all large-scale innovations – continue to be shaped by, and shape, the cultures of biopolitics and identity, and not just in the US. It is inevitable that executive politics and science work with each other, even if, as Chris Mooney claims with considerable evidence, this particular period was an extreme case of distortion (ibid.).

In 1973, when Jane Roe challenged in the US Supreme Court on behalf of pregnant women of Texas, for the right of women to an abortion (*Roe v Wade*, state cases) a precedent was set for the introduction of abortion in the US, with trimester guidelines then to follow. Then in 1992, the *Planned Parenthood v Casey* decision in the Supreme Court invalidated *Roe v Wade* of 1973. Elements remained however, the main issue being a change on the issue of 'foetal viability'.

Diamond v Chakrabarty in the US Supreme Court was the seminal moment for the protecting of the patenting of living organisms, reflecting a tendency toward an ideal of the stewardship of nature, an American form of conservationist environmentalism. Leo Marx's (2000)[1964] *The Machine in the Garden* has painted a picture of an American pastoral where a technological object can co-exist easily with the human steward of Nature. This can be seen as a sociological part-explanation for the US relative acceptance of genetic modification in agriculture (Gaskell *et al.*, 2000). While it was a particular historical route that brought the United States to build such a garden machine, what is interesting in this contradictory image is the type of hybrid deemed necessary for future technological development, a constant changing mix of the old ideas of Nature and Culture (Latour, 2004a). When it comes to regulation of RGTs however, it is also interesting that the US has not fully legislated for RGTs, allowing a liberal agenda of stem cell production by private funding while keeping federal funds to the whims of ideology of government at a particular time, or indeed of individual states. This has occurred despite high-profile campaigning – indeed bio-oration – efforts in support of stem cell research funding by Michael J. Fox, Ron Reagan Jr. and the late Christopher Reeves, each of whom appear in a film used in the workshop sessions in this study, *If...Cloning Could Cure Us* (2004a). It must also be noted that genetic screening is a costly and prohibitive procedure in the States, costing anything over €30,000 (Beacon CARE Fertility, personal correspondence).

Germany and Britain, on the other hand, have had different sets of circumstances. The birth of Louise Brown, the first 'test tube baby' in Britain in 1978 was the common paradigmatic case, the origin myth. It seems interesting now, looking back, how unsettling this event was for many as a news story. IVF is now a common procedure, but then there was a novelty to this lab facilitation. Genetic screening raises many other issues and should not, of course, be tied too strongly to abortion. We will see in Chapter 7 the distinctions young people make on RGTs. For Germany, of course, the spectre of eugenics is all too real, and made all the more so in Arendt (2006 [1963]) in how she describes how a 'banal evil' can creep into society, however difficult it may be to place the plight of a couple wanting a child free of disease alongside the methodologies of the Third Reich (Habermas, 2003). As Jasanoff (2005) reports, and we examine again in our genetic futures chapter (7), there is a certain 'sanctity', albeit secular, afforded to embryos in Germany under Article 218 stGB (abortion) (ibid. 24) (Embryo Protection Act, 1990; Jasanoff, 2005: 158). There is on-going debate in Germany about the use of non-pluripotent (non-programmable) stem cells and a recent court ruling in favour of PGD opening the door for its more widespread use. In the UK, a special expert panel on science and ethics was convened, leading to the 1987 White Paper on Human Fertilisation and Embryology. The British Parliament passed the Human Cloning Act in 2001 (Jasanoff, 2005: 199).

Other European countries also have legislation in place for RGTs (ibid.). Ireland is one of the last to do so. Individuals and couples across Europe have used the European Court of Human Rights to gain reproductive freedom, access to information, or indeed access or donation of gametes. There are some legal conundrums ahead.

Biotechnology has become an area of heightened political importance in China in line with the One Child Policy (Klein, 2010) – this does not escape the attention of some of the young students who participated in workshops described later in this book. While there is a ban in China on human reproductive cloning, therapeutic and research procedures are permitted on the human embryo (ibid.). India has banned therapeutic cloning, but in line with many cultures globally, sex selection of embryos is common where there is either a social stigma or bloodline implications to having all-female progeny. KK Women's and Children's Hospital (2013) in Singapore claims to have carried out the first IVF procedure in Asia.

Countries of Roman Catholic tradition in Europe demonstrate the influence of religious institutions in regulation. There is a more acute case following recent abortion legislation in Ireland, enacting a twenty-year-old court ruling, spurred on partly in response to global media coverage of the death of Savita Halappanavar having been refused an abortion in an Irish hospital and the political and elite media obsession with abortion that followed. The debates about reproductive technology legislation are inevitable in Ireland where there is a legislative vacuum regarding this. Ireland's is a unique case: Article 40.3.3o of the Irish Constitution (Bunreacht na hÉireann, 2004) refers to the embryo, euphemistically and quasi-legally, as the 'unborn'. The unborn is the ultimate hybrid in its discursive complexity (Gieryn, 1983; Latour, 1993). There are vague guidelines in Ireland regarding the age limit for genetic testing under Medical Council Guidelines. In recent years, individuals who undergo testing in fertility clinics, two of which were visited as part of this study, the public National Centre for Medical Genetics and Beacon CARE Fertility, receive counselling on the consequences of their actions. On the other hand, Italy has regulation in place effectively banning PGD, but cases have occurred through court order. Both Italy and the Spanish Fertility Society (SEF) also prohibit germline therapies and surrogacy. Spain has in place agreements with private institutions on the donation of gametes (Centre for

Bioethics and Human Dignity, 2013). There is debate in Spain since PGD's introduction about the possibility of this being a type of eugenics (Rodriquez, 2012; Vivanco *et al.*, 2010).

A structural power can take subtle command and we may not see it. The major example from history is eugenics. Eugenics was once pervasive across the political and connected higher education landscapes of developed countries. While we agonised over the differences between therapy and enhancement in the first chapter, those differences mattered less in the nineteenth century. However, the difference is the loci of control: we consider what autonomous individuals can be empowered or allowed do in contemporary times, but in those times it meant governance of the ill, the feeble-minded, the criminal and the weak.

The voices that contribute to how we today know biology was partially impacted upon by personalities, key influencers. They are referred to here as 'bio-orators', a tradition of communicating biosciences that changes somewhat as it moves historically within the paradigms of biology from Huxley to Uexküll to Cox. They have set the discourses for how we *see life being discussed*, of particular relevance to RGTs.

The history – accepted history – of winners has needed its orators. Science needs progress; progress needs politics, and the Politics of an era needs its Political speakers to push action forward. If we look at the timeline of biology in history, Table 2.1 shows biology's dangerous ideas, to paraphrase Daniel Dennett.

They are shifts in structure connected to a new knowledge, a new application in biology that impacts on our ideas of identity or indeed this co-constructed knowledge in society. Let us look through the intellectual ideas that were threaded between these epoch-defining events in biology and biotechnology. This was a biopolitics that shaped, and was shaped by, nineteenth- and early twentieth-century political thought on bodies and populations and the strategies of control. Bio-oratory emerged from modest to high-powered speeches and rhetorical lectures that have reverberated, an integral part of how we know biological, and particularly genetic, concepts. They include Paley, Huxley and Galton in Britain, as already mentioned.

Eugenic thinking has a long history. As we think about the very foundations of republican democracy in the US, France and Ireland, we may care to omit Plato's reference to eugenics in *The Republic*, '....there will be considerable scope for ..."right use" in marriage and procreation...[to] mate the best of our men

Table 2.1 A timeline of biology's potential threat to identity and society

1800s	Darwin's *Origin of the Species*
1800s–1945	Eugenics programmes
1950s	Double Helix of DNA
1970s	Sociobiology
1980s	Biotechnology
1990s	Cloning mammals
2000s	Human Genome Project

with the best of our women' (Plato, 1987 [360–380 BC], 459d: 180–1). It is an immunitary urge (Esposito, 2008); however, cleansed of its life politics, the societal value morphs into discourses of health, industry and agriculture in the classroom, a markedly positive force for good in society. The messiness of controversy removed, the taboo sequestered. Some influential orators of the biosciences such as Huxley, and others in the twentieth century, ultimately brought bio-control down dangerous routes politically.

So who are our bio-orators, those in history who have shouted loudly about what we know of biology, control and the body? When Boyle argued with Hobbes over the implications of the air-pump, it was the start of a 'modern' battleground of speech and ideas about Nature and artifice, moving downwards from structure that Hobbes envisages as a rational Leviathan in sovereignty, way down past Boyles' curious introduction to the world (winning history tells us) of the agency and magic of things before a crowd (Shapin and Schafer, 1985). Boyle, as well as being a brilliant 'chymist', perhaps had charisma and luck, which our bio-orators also need, with a little media hype. For those from the latter part of twentieth century onwards, there is media transference of the power of the personality of the speaker onto the enterprise of science, which in turn has, since Hobbes and Boyle when it was Natural Philosophy, contributed to the social order.

We could start with **Thomas Henry Huxley,** Darwin's great defender during his lifetime, known as 'Darwin's Bulldog'. He wrote an extraordinary long and complimentary review of *On The Origin of the Species* for the 26 December 1859 edition of *The Times*. In his famous 1860 debate with Bishop Wilberforce at the British Association meeting at Oxford, the line that was attributed to Huxley in the press at time, when pressed by Wilberforce whether on the maternal or paternal grandfather's side he had descended from an ape, was: 'I would rather be descended from an ape than a bishop'. Whatever was said, some incident of hasty discourse occurred in the room by common agreement, but there are many versions of the tale. A woman supposedly fainted, and by other journalistic counts, Huxley was even said to have been aloof, looking sickly, when he allegedly uttered the phrase (Lucas, 1979). Perhaps it was not such a brilliant speech after all. It matters not; it is now legend, part of history hailed for many as a battle won in the war between science and religion in the nineteenth century and another step in the popularisation of Darwinism (Huxley, 2012[1903]). The debate that followed at the British Association event two days later was recalled by his son Leonard in praise of its brilliant oration: 'It was, moreover, not a contest of bare fact or abstract assertion, but a combat of wit between the individuals, spiced with the personal element which appeals to one of the strongest instincts of every large audience' (ibid. 262). Yet Galton took this on, following Malthus, and indeed the starting of control of the populations began and manifested in the fiction of HG Wells as well as grandson Aldous Huxley's *Brave New World*.

As bio-orators, we could also mention the 'social science side' of **Jakob von Uexküll, Georges Canguilhem** and his student **Michel Foucault**. Uexküll's views on life were profoundly political and influenced education and life politics.

A vitalist, he developed the earliest work in the field that became known as biosemiotics. Uexküll introduced the term *umwelt*, for him meaning a close environment to an organism influenced by and influencing the organism (even ticks have a surrounding subject sphere), a term adopted into the world of philosophy and the social sciences as well as the natural sciences. The *umwelt* of Uexküll splits into philosophical-minded biology and sociology. But it was the supervisor of Foucault who does the oratory here. Canguilhem (1992) argued in his 1947 lecture *Machine and Organism* that there was something different about how biology and technology of the twentieth century have reciprocal action. His rhetoric, having seen the codes and communicative objects of Crick and Watson, was that life can be explained through mechanical analogies. Conversely, objects and technologies of our Progress are themselves extensions of life. The matter-subject ideas have evolved down one strange pathway in the 'speculative realism' of Jane Bennet, Graham Harman, Quentin Meillassoux and their ilk. The relationships between geographic environment, human, and non-human has been explored in contemporary STS by Thrift (2005), organisms and things extending outwards, learning and teaching oneself/itself in the world, forming while informing. And, as already mentioned, Foucault, most eminent of French philosophers, took up the challenge as Canguilhem's student. His position of influence on the study of body-as-state and the bodies within/below a state is peerless. Foucault's famous revolutionary lectures of 1976 as a social scientist talking biology and the body introduced the world to governmentality and governance through the body's positioning in social spaces, governance that has a bit of the government about it. And it is to governmentality and loci of control with RGTs that we keep coming back.

Uexküll's oration to the Nazis may not be so illustrious. Harrington (1996) describes the scene where Uexküll attempts to point out to the assembled National Socialists the errors of their ways at Nietzsche House (also present was Nietzsche's sister), using body-state biopolitical analogies, how universities were the vital organs of the state and needed to be protected with minimum of Jewish 'immunisation' (ibid. 70). The lecture was badly received by its audience, interrupted by Socialism Democrat guests for veering off the evening topic (legal philosophy), yet the idea of body-state and *immunitas* took hold in Third Reich ideology. I feel cause to mention at this stage how insensitive it seems to introduce Nazism to the topic of prospective parents desperately wanting a healthy child, for the child's sake. But this is the point, made clear throughout this book – we must also keep responsibility for the larger picture, the structural issues, such as concerns over state decisions of future technologies governing genetic characteristics.

Henri Bergson is associated for many with his debate with Albert Einstein on the Theory of Relativity, and this indeed was the proof of a man unafraid to challenge the best. Bergson came out second best in that debate by common agreement, and in fact Latour and associates staged an enacted 're-match' to give Bergson his opportunity in history to show how we may think differently about scientific theories as public things (Latour *et al.*, 2010). But I would also like to mention the Gifford Lectures in Edinburgh University, 1914, as it was here that Bergson voiced concern for an ordered universe that scientists felt at the time did not require something fundamental in personality, multiplicity and community. The understanding of science, he also said, required will (Bergson, 1914). Of course this concept has cast dark shadows across the twentieth century. However, it is worth noting that these speeches were also a challenge to both church and science. It is here that Deleuze draws on bodies and 'rhizomes'. Foucault and Deleuze are direct ancestors to the Italian fringe of Negri, Esposito and Agamben, a section of contemporary philosophical thought that are the flag-bearers for how a new thought about biology teaching should come about. This is emerging as a new biopolitical philosophy, influenced by Foucault and taking from the deeper meaning in Greek etymology of the *bíos*, the political forms of life that strive to live, that exist for a good life, rather than bare life *zőë* (Agamben, 2000; Esposito, 2008), a life that is less-than-human, beneath our societal protection. An extreme form of *zőë* is the 'state of exception', how a state can control a body and vest it with such symbolic criminality that it is deemed outside state or international laws, susceptible to long periods of solitary confinement, torture and lawless death. For Italian philosopher Roberto Esposito (2008), *bíos* still has its roots in a negative declension of biopolitics where there is an immunity reflex in biology with further sequestering, a weeding out of undesirables. RGTs can be argued to do something similar for the human race.

Whatever about Canguilhem's dedication to oration in the academy, or Huxley's legend of showmanship and the putdown, Bergson was a public intellectual, reputed to have created the first traffic jam in New York at his Columbia University lectures. Even if his ideas of *élan vital* and consciousness are often derided, and hardly popular wisdom (Gamson, 1992, and see Chapter 6), his style and legend have grown, helped in academia by Gilles Deleuze and by a major public engagement effort by the *New York Times*.

We cannot simply stack the rhetoric of Huxley against Bergson: they were different eras, different worlds. The views of some these defenders of humanity would appear politically incorrect today. This further adds import to studying bioscience in historical context not as an anachronism but complete with its messy social origins. These figures in history, albeit as figureheads and nodes in the network of biology knowledge, of biology for change and thus life politics, could change science education. It is a consideration of both sociology and biology as controlling, as dominant structures in history of thought. Eugenicists were bio-orators – how do we, the 'we' that stand with educators and young people, respond to these out-dated ideas today? Are there echoes of positive eugenics in choosing and immunisation of future bodies?

Habermas, as we mentioned in Chapter 1, is a voice against eugenics. Although perhaps not the best orator, his battles alongside Dworkin and against **Peter Sloterdijk** were primetime TV material. This was Germany, and the giant shadow of the *Ubermënsch* still stands over the political system and public discourse.

Nietzsche's *Ubermënsch* was a direct descendent of Hobbes' Leviathan (and maybe even Stan Lee's Superman), a body politic, but the Nazi regime brought this concept to an evil conclusion. It is here also that the potential for Heidegger *thanatopolitics* emerges. There are no rhetorical or pedagogical equivalents in the Anglophone world to continental media debates of the *feuilletons* where intellectuals can exchange robust views for the good of future humanity. The links with breeding, even the uncomfortably active term 'selection' when describing future genes were invoked, and, like Esposito, Sloterdijk saw an immunitary, exclusionary thinking in the acceptance of PGD and the implied right people feel they had to future body perfection. One issue is the elitism of these speeches, and public and intellectual reaction to them, unlike the multitudes who turned up for Henri Bergson in New York.

When **Erwin Schrödinger** arrived in Ireland in 1939, the chemistry and physics of life was in its infancy. The Dublin Institute for Advanced Studies was, by its nature, elitist. However, during the Second World War, known in that downplaying Irish way as 'the Emergency' in neutral Ireland, Schrödinger caused a minor stir, the cause of parliament debate over his stay in the first place (a pragmatic reason for his stay in Ireland was his flight from the Nazis in his native Austria), and a familiar eccentric cycling from his home in Clontarf to DIAS in Merrion Square, or in *lederhosen* in his country excursions (Moore, 1994). He debated God, physics and morality with his friendly foe, Monsignor Paddy Brown, a mathematics teacher at St. Patrick's College, Maynooth, who had serious objections about the published *What Is Life?*.

Schrödinger's influence on Watson and Crick is well established. Watson has become a controversial figure, with accusations of racism (Blue, 2007); he certainly espouses publicly a biological deterministic worldview. Once recombinant DNA techniques were developed in 1973, Watson helped lead the way in the 1970s' and 1980s' biotech 'revolution' in recombinant DNA and biotechnology. In the 1990s Watson and the Wellcome Trust-funded Sanger Institute went head-to-head with **J. Craig Venter** in the race to be first to map the Human Genome Project. It was a contentious struggle – should not a publicly derived, social contract with science mean the efforts and products remain public? This was to be the finest technological advance of its time. There was a rush in the end to get the project finished by the year 2000, symbolically putting a line between one epoch and another (at least one voice from HGP, John Sulston, was loudly warning of the deterministic dangers of the HGP). As Bryan Appleyard (2004) noted, DNA is a prominent visual icon for the twenty-first century. There were benefits trumpeted around bioeconomies of the HGP; Sanger discovered the BRCA-2 gene, the presence of which increases the risk of breast cancer.

And although voice is the key to oration, the visual aspect was important here. We begin to see a difference here in our bio-orators, in that our 'orators' are moving from voice alone. In line with how Giddens would see the key influencers in identity construction, so celebrity, that is ubiquity on media platforms, does the work of persuasion. According to Cavarero (2005), there is a growing distinction between voice as performed speech act and voice as semantic democratic ideal. It is form and style rather than composition at this level, the mediated technology of TED and PowerPoint that gives us the 'stuff' with which to compose. Craig Venter fits the bill as a business guru with a vision and a good presentation, rather than the older, more political notion of an inspiring, intellectual lecturer, like Bergson. Media and mediation is key and the visual medium most important of all – YouTube, infographics, soundbytes across a screen become the new bio-'oratory'. For Cavarero (ibid.), a new voice of vocal chords, rather than semantics, now supplants this, the voice as voice-over. There is rhetorical depth in the documentary, for example, the new power of celebrity as a different type of public intellectual, where the intellect is not tied to the dusty offices of the academy.

Televisual rhetoric for the biosciences is a key mediation, *and* a witnessing. Let us remain with that famous press conference announcement in 2000. Two continents were represented by President **Bill Clinton** and Prime Minister **Tony Blair** (the *de rigueur* satellite link-up) and the Republic of Science represented by Vietnam medical vet, J. Craig Venter and his whole-genome 'shotgun' method (YouTube, 2000) and Francis Collin. While it is an open question what to do with the mapping of the human genome, the healthcare sector is one area where is it hoped that genetic markers can be matched to diseases and treatments. Two males and three females of different ethnicity contributed to the whole genome mapping. It is the perfect study for STS – five genomes and the whole human race on one map (biologist Bruce Lewontin, something of a bio-orator himself, is among the sceptics when it comes to using HGP to identify disease). At the White House, as the US anthem is played, Clinton, Venter and Francis Collins, the HGP lead, step out to the press conference, Clinton flanked by the other two. A 'momentous year...this day for the ages', Clinton says. He welcomes the ambassadors and lead HGP scientists of the world (France, Germany, Japan, China, as well as US and UK). Clinton, in opening comments, discusses significance of the room, how two centuries earlier when Democrat-Republican origin myth President Thomas Jefferson announced a map of the Western frontier. Geographies of the past and future frontiers. A nod to Watson, also in the room, who seemed to have British understatement in his initial paper. Clinton reads the famous conclusion from the seminal paper: 'This structure has novel features which are of considerable biological interest'. Clinton compares the venture to Galileo, and echoing his careful rhetoric to the Pope, then says we are 'Learning the language with which God created life', taking in the religious constituencies. There is allusion to 'healthy competition', meaning the public and private race – no hint of controversy here – and how this is only the beginning. There is obvious *logos*, to ensure the message is clear about the acceptance of the presence of biotech companies, the source of much controversy in the project. But Clinton's speech ensures human values remain: this must be 'used for the good', not to 'stigmatise or discriminate'; among the inequalities of the US. Suddenly HGP is the great leveller.

Then Tony Blair patches in via video link from Britain, with Sanger and Wellcome Trust representatives. Another big technology of the millennium.

Clinton, ever the rhetorician, mentions Blair's newborn son, Leo, born on this landmark year. 'His life expectancy has just gone up by 25 years.' It is interesting that Blair plays the role of the European bioethical sceptic; he raises the risks immediately, how we must 'develop the possibilities and thwart the dangers'. On political themes, gratitude is immediately granted to Clinton for other joint work on the Northern Irish Peace Process. Blair returns praise to the US side, and the other seemingly minor national partners in this announcement, but also is careful to mention NGOs. He refers to the 'superhuman task', provides downplay in that it is 'all but complete'. The British watchdog voice is invoked, as 'with power comes responsibility', that ethical and moral questions raised by this research is 'common property' (implicitly private and public). 'For the common good' and 'all human kind' are mentioned three times each in Blair's speech. Like Clinton, there is unification, the 'shared values'.

Both Blair and Clinton's 'clear, defined goal' invokes multi-nationals and states working together and the possibility of different jurisdictions for ELSI mirroring the development of the HGP itself. With this rhetoric, science and ELSI are split, but operate along the same global composition. Gaskell, Bauer and their colleagues were quite astute in their analysis of biotechnology as a major political and economic force in the late twentieth century, focusing mainly in relations – and tensions – between US and EU (Torgersen *et al.*, 2002). (See Jasanoff (2005) from US research perspective.) If we look at another genetic global composition, GMOs for example, what is striking historically is the contradictory, two-pronged nature of the global approach, perhaps most contradictory in Europe: on the one hand, the go-ahead is there for competitive national and multi-national growth in the biotech sector of the markets. On the other hand, there is the Precautionary Principle of health and safety checks (ibid. 2002). These contradictions are indicative of the Beck/Giddens convergence of risk society theses through reflexive modernisation of institutions (Beck, 1994; Giddens, 1994; Lash, 1994). There is a link to Giddens in Blair's presence, as his 'Third Way' structure of political and educational thought was adopted into the political philosophy of New Labour.

At the HGP press conference, Collins then speaks and quotes another Pope: 'Know then thyself, presume not God to scan'. Other framing devices and metaphors were left with us that day: 'reading our own instruction book', 'the book of life'., 'reading the language of the genome', all literary, each meaning supposed open access to all scientists. There was added *pathos* in Collins's sister's death from breast cancer. When Venter speaks, he rhetorically places himself with the pantheon of scientists of life (including his rival Watson). Venter's phrase 'Life reduced to its very essence' becomes the popular wisdom for genetic study thereafter in the new millennium. 'Wise use' and 'scientific literacy' will prevent 'discrimination'. 'Poets and philosophers', Venter says, answering the critics who cry disenchantment, and in his doing so, wraps up the press conference with these carefully chosen words, 'will have inspiration for the millennia'.

This is demonstration (Barry, 2001; Callon, 2004) again, the feeding off the media-need. They are political performances necessary for systems of governance, regulation and control of the processes of 'bio' – the information and ways of practice, the styles of thought. To see this new biopolitics as something that will require more global/national governance, the education system needs to see this, but also to engage with the cultural oddities of thinking about 'future bodies'. The cultural and communicative becomes a *vital* link between classroom and governance. Mode-2 displays realities of complexities: such as the difficulties of the social contract of Rousseau, Hobbes; Gibbons and his co-author Helga Nowotny, who became Director of the European Research Council, also provide their version of a new social contract (Gibbons, 1999). But, according to the amodern paradigm described by Latour in Chapter 1, 'science' and 'society' are not separate entities; even Giddens' late modernity would find the boundaries difficult to identify.

STS research and practice have provided plenty of evidence of a networked, Mode-2 science, the Latourian parliament idea is difficult to find in the foundations of science education. We essentially need more critique within science, this self-confrontation that is a publicly visible (witnessed) as part of reflexive modernisation. It filters down from our governance, from our speakers, particularly if they are established as public intellectuals. This self-confrontation of science is more visible in some countries than others. While Britain has the Royal Society, the British Association and the Wellcome Trust carrying on traditions of debate-oriented public engagement, in France, Louis XIV established the French Academy of Sciences in 1666. The Spanish Royal Academy was founded in 1847, although as stated elsewhere the Civil War years stagnated scientific progress. Yet as Trench (2001) observes, Ireland, another Catholic country, is missing this intellectualisation of the sciences.

During the ramping-up years of bioscience funding, **James Watson** was an occasional visitor to Ireland, then working with Cold Spring Harbour, before being let go for one too many controversial remarks. In the early 2000s, institutions like the Smurfit Institute for Genetics, TCD and the Irish Research Scientists Association were involved in inviting Watson to increase local prestige for genetic sciences. Ireland's so-called 'turn to science' began around this time. The 1996 White Paper on Science established the framework for an Irish advisory body in science, a Technology Foresight exercise, and an inter-departmental group set up on biotechnology and the eventual investment of €7.1 million of Government funds on biotechnology and IT, the then twin pillars of Irish science and technology, with some university infrastructural support funding from Chuck Feeney's Atlantic Philanthropies. A public engagement remit was set up with a group called Biotechnology Ireland, but was ultimately short-lived – a communication website established in 2000 was quietly taken down by 2003. We will look later in the chapter at other 'outreach' initiatives.

But it was a start for biotech in Ireland, laying the groundwork for the life sciences as economy-boosting activities. Since the strange days of President De Valera and Erwin Schrödinger, Ireland's reputation in the sciences has come into its own; the twenty-first century turn to science brought this further into focus, often pushing above its weight in terms of European and commercial funding of

STI. But its intellectual import, in terms of collisions of ideas, has been downplayed in favour of larger narratives of an Ireland of 'saints and scholars', its religious foreign missionaries and literature traditions (Trench, 2001).

For a contemporary, televised twenty-first-century audience, I could present David Attenborough for the medium of TV, and Richard Dawkins for arguing persuasively his dangerous idea of selfish genes. However, **Brian Cox** is a speaking superstar, having taken part in many medal awards lectures. Arguably the biggest science celebrity in the UK, his presenting of the series the *Wonders of Life* marks a disciplinary turning point for him into the realms of biology. Cox has youth-identification credentials, formerly being the keyboard player for pop band D:ream, albeit not a cool band, which disbanded a year before his PhD in astrophysics at the University of Manchester. While there may be a visual and conceptual reductionism of the *Wonders of Life*, it is the celebrity expertise of Cox casting a physicist's eye on biology that has captured wide appeal, a lineage of interdisciplinarity in biology that stretches back to Schrödinger. Cox, however, is in the business of explaining and disseminating, not engaging in political debate. In fact, with others on the Caged Monkey Circuit such as Robin Ince, Cox has made the mistake of attempting to force apart Science and Politics. It could be argued with Dawkins, EO Wilson, Lewis Wolpert and others, the tradition of scientism is kept alive, with public support. We could also consider here Robert Winston, a politician and broadcaster, who just happens to work in reproductive medicine (both he and Cox are contracted with after dinner speaker agency JLA).

The political of Mode-2 and institutional self-confrontation leads to activism, such as Genetic Interest Group in the UK and the Genetic Rare Disorder Group in Ireland. Barbara, Mark and Annie in *The Gift* would have had dealings with such political pressure/lobby groups at some point. **Christopher Reeves**, a Parkinson's disease sufferer who was a vigorous campaigner in his latter years for federal US funding on stem cell research, appears briefly in *If...* He was a celebrity embodiment of anti-ableism for individuals trapped in damaged bodies, the actor who played Superman in the movies of the late 1970s and 1980s. There is considerable irony that this Hollywood *Ubermensch* should spend the last part of his life in a wheelchair. Reeves, like other public figures such as Ron Reagan Jr. and Michael J. Fox are referenced in tiny clips in *If ...Cloning Could Cure Us.* Like Andrew, the wheelchair-bound male lead in that movie, he wants a different life, a more mobile one like the one he had before. This is in contrast to the acceptance of fate by FRDA sufferers in *The Gift.* We do not see Annie, the young girl who develops FRDA, in her later years, but we can see her friends in wheelchairs, part of the advocacy group she joins, who are involved in protesting to get her brother tested for the disease. And we know from expert systems of knowledge the likely fate of FRDAs sufferers. For Reeves, of course, there may have come a point of transference, where his hope was for the treatment of future constituencies, as time was running out for his own plight.

While here are uncomfortable issues with our bio-orators, the main issue is more obvious. They are all men. And so we come to perhaps the less well known, who, nonetheless have their place in RGT 'future histories'. The final two bio-orators are surprising, not least because both are female.

The landmark *Planned Parenthood v Casey* Supreme Court decision in the US in 1992 invalidated previous abortion legislation from *Roe v Wade*, while keeping certain elements such as 'foetal viability'. The new ruling represented a stronger position for women's rights. **Kathryn Kolbert**, the *Planned Parenthood v Casey* lawyer and ex-journalist defending elements of *Roe v Wade*, made a staunch argument for women's reproductive rights in 'law rather than the proclivities of individuals' (C-Spanvideo.org, 1992). Without these, she argued, women of the US would be forced into back-street, unregulated abortions. It was a defining year in the US in many ways. Abortion became a presidential election issue, and a conservative vote probably carried Bill Clinton along, following his comment at a press conference before the Supreme Court verdict that the right to choose was 'hanging by a thread' (Devins, 2009: 1321, quoting Schneider, 1992). That thread was Kolbert. Political media commentators are quick to identify single defining moments and utterances on media that swing opinion or votes. We cannot say this one bit of media talk won it for Clinton the following year, but while Kolbert's speech has been widely credited as the reason *Planned Parenthood* was passed with a wider ranging win for women and reproductive politics, Clinton's win was for Politics but with less longevity; Clinton got four more years in the Oval Office, while Kolbert changed the lives of many American women.

Finally we have **Joanne O'Riordain** who said, in the way only a child can say: 'Build me a robot!' It was a challenge laid down at the UN International Telecommunication Union Conference 2012, a challenge since taken up by Trinity College Dublin. Joanne has total amelia, which means she was born without limbs. She has overcome remarkable odds to enjoy what she hates to call 'normalcy'. All she asked for at the UN speech was for someone to build a simplified version of an entity straight out of science fiction to pick up the things she drops. Before that, Joanne challenged the Irish Government to reverse swingeing cuts to disability services. She may or may not have helped the cause, but there was a slight easing of austerity in the disability sector following the speech. The best role models for young people are young people. Malala Yousafzai was also witness to terrible misogynistic prejudice by the Talaban that almost ended her life. Malala was shot for attending school, disallowed for gender. She has campaigned powerfully for education and women's rights at the age of sixteen. The power of these two voice in the matrices and practices of on-going technological and biopolitical change. Joanne, more than Christopher Reeves, represents the anti-ableism agenda, the cultural bias against the disabled (Fuller, 2011; Wolbring, 2008).

The cultural salience of the 'gene for'

The inclination to have a cause linked to one gene, positive or negative (Nelkin and Lindee, 2004) has media prevalence and cultural salience. Easy genetic explanation leads to genetic essentialism, what Nelkin (ibid.) and others have called a 'gene for' phenomenon. The Dawkins/Gould conceptual–historical battles of the 1980s and 90s contained this essentialism at its core, particularly on the Dawkins side.

The discourse of biology is moving from the nuclear material of genetics out into the cytoplasm (Rose, 2007), but the ideology is the same – diagnosis and prescription leading to control and immunisation. Biomedicine with genetic knowledge can be set up to decrease or even eradicate 'differences coded as defects' (ibid. 55). More than genetic essentialism, representing the molecularisation (ibid.) of Ireland, life politics emerges on the highest European political level, also enacted at every quarter as the flow of information and genetic data; this proteonomic power supposedly moves from lab to bench, from the diagnostic to the therapeutic. There is political heat and immunitary action (Esposito, 2008) already implied by the screening practice of screening cells that have a successful transfect – that is selecting those cells with newly formed pieces of DNA from a molecular cloning procedure. Bloodlines are replaced by cell lines. A cell line known as HeLa is often characterised as embodying Henrietta Lacks, although of course, how can it be her? Is it not her cell line only? Yes, but HeLa has become more, and we forget the human suffering of a life, made real in the popular book *The Immortal Life of Henrietta Lacks* (Skloot, 2010).

Dorothy Nelkin describes the gene as a 'cultural icon' in *The DNA Mystique* (Nelkin and Lindee, 2004). Its image has permeated our conversations in recent decades, increasingly becoming associated with new imaginings of fertility, family, animal–human chimeras, and body enhancement. Nelkin and Lindee (ibid.) describe how genetic issues such as cloning and PGD are salient in the US because of that country's obsession with biological identity. The same could be argued of Ireland, given its long fascination with a cultural discourse of distinct Irish heritage (O'Mahony and Delanty, 1998) and nation-state discourse of the essence of 'Irishness' (Hill *et al.*, 2000). There is also evidence of an emerging scientific discourse on 'Celticness' as part of late modernity's self-identity (McKie, 2007; Battles, 2002).

As Nelkin and Lindee (2004) describe it in *The DNA Mystique*, 'gene talk' is part of our everyday discourse, going far beyond the science. We hear it said that success is in a company's 'DNA'. We hear terms like 'genetic' and 'genes' in politics, film, theatre, art, fashion, newspapers, fiction, games, graphic novels, and everyday chat. Writers such as Dorothy Nelkin, Susanna Hornig Priest, and Stephen and Hilary Rose see values propagated by both media and popular culture as a *scientism*, a ready acceptance of genetic explanation without question. Genes are used as bits of information, using the ontology of computer design and communications. We have used engineering, IT and process control metaphors for life, in terms such as 'code', 'messages', 'cellular communication', 'pathways', 'evolutionary flows'.

The STS scholars cited here have a common concern for the salience of the 'gene for' frame. As Appleyard (1999) describes it, the DNA double-helical structure looms large in the background as a type of postmodern icon, and indeed is used extensively in both biotech and non-biotech branding. Yet technological applications have deeper salience in the current Irish political and industrial system. Labelling something genetic clinicalises, it creates an institutional exclusion of the moral, existential and social of RGTs from educational and public discourse in a genocentric society (ibid.). Nelkin and Lindee's (2004) view is that a broader genetic essentialism still prevails across *all* cultures of modernity, where genetic explanations are offered uncritically – sometimes based on a single gene – and this phenomenon is potentially on the increase.

Science, state and schooling: the biosciences' impact on twentieth- and twenty-first-century education in Europe and the US

Ireland became an independent, sovereign Republic, a representational polity that upheld the dignity and safety of the collective of individuals in that state. We cannot and should not go back to the 1930s constitution, but can we see a new constitution that has a spirit of Latour and Giddens? Between 2010 and 2013, sovereignty was suspended. Ireland was a country bailed out financially by the International Monetary Fund, the EU and European Central Bank. Ireland is currently debating its futures now, its status as a Republic, facilitating a National Constitutional Convention. Can we become a new Republic? What does this mean for interactions with global bioscience governance and the neoliberalist tendencies of RGTs? Giddens tells us perhaps we need to conceptually give up on the Sovereign eventually in a globalised process. Foucault has already separated biopolitics from state. Also Hobbes' *Leviathan*, visited throughout this book, can tell us much, still, about the commonalities between scientist and politician. Within these dilemmas, Ireland has had a difficult, intense relationship with concepts of scientific view of life.

The new Republic's formal education system consisted of 'national' (primary) and secondary schools, with standardised state exams set for equality, instrumentally creating a level playing field (the Irish Leaving Certificate at senior level is the curriculum level at which the school sessions in this book are pitched. One particular textbook will be examined in Chapter 5). Founding figures of the Irish state such as Pádraig Pearse brought a certain wilderness romanticism to the earliest curriculum on biology education in the early twentieth century, the 'nature study' of beauty, of which the emerging agricultural state would have control over their own land and livestock which grazed upon it (Walsh, 2007). It was an immersion in Nature, but of course such fieldwork has also provided sequestration, taking the technical administration out of science, which is what science became in the twentieth century.

At other points in this book there is subtle demonstration that a religion versus science debate gets us nowhere but polarised. And so it is the case here when historians describe simplistic anti-science bias in religions, and use the Catholic

Church as an example. In fact, as Fuller (2011) shows, there was a certain Protestant asceticism of human subject and Providence that found it difficult to swallow Darwinism. Irish Catholicism did not object so much; perhaps animalism is not one of those red alert issues for the Church. Whyte (1999) sees little evidence of the epistemological thesis of Protestant Ascendency that posits that evolution and other existential concepts were discouraged by Catholic clergy, while Duddy (2002, cited by Trench, 2001) maintains that Boyle, that originator of empirical science as social order, was a uniquely Irish thinker.

For Ireland, a newly independent state, it was Britain itself which was the sequestered idea, the enemy which 'provided science' in the histories of Darwin, Newton, Faraday and the Royal Society, even the Royal Dublin Show, even though the latter was instrumental in establishing the Botanic Gardens and the National Museum (op. cit.). The new nation ideology sequestered where it could any such links. Gaelic culture, national history, the Irish language, sports, arts, civics and commerce took precedence as pedagogical focus over British historical processes of science. As remarked at the beginning of the chapter however, sectarian educational practices in Ireland that favoured Protestants did exist, the 'gate-keeper' thesis developed by Krishna (1992) when referring to Indian science and colonialism, and so it would be expected that scientists and science emerged from these classes.

By way of contrast, if we look at another Catholic country around the same time, the educationalist Renzo Titone (1958) describes a positivist/spiritualist split in Italian education. While eugenics policies informed education policy generally, the common trend among nations where education had that same immunitary task of racial and social cleanliness, concentrating on the elites as a link between descriptive biology and active sociology (Cassata, 2011). These philosophies were part of the shaping of the *Legge Casati* (Casati Act) for *Scuola secondaria di secondo grado* (secondary school) in Italy.

In Spain, where Averroes first brought Greek mathematics from Moorish intelligentia, Franco's regime after the Civil War effectively stalled scientific progress. Only Peregrín Casanova's university textbook commemorating the fifty-year anniversary of *Origin of the Species* in Spain remains as a rhetorical beacon for Darwinism education within this period (Barberá *et al.*, 1999).

Certainly the Catholic doctrine *is* a likely reason why faith has not been part of this sequestration in the curriculum strategy, in the manner of science. The sequestration referred to will pop up frequently, a sequestration of experience where a mediated (in this case a simplified understanding) of science and technology continues the Modern journey of removing what are common-sense elements of existential doubt and complexity of feeling (Giddens, 1991). Are we seeing a sea change or gradual change away from a Catholic to a scientific Ireland? We cannot say without evidence that we live in a post-catholic Ireland. There are many complexities between religion, science education, ethos, ideology and a *zeitgeist* of a people; however, a common theme is the hybridity, as Gieryn (1983) would call it, between an objective bioscience, historical processes or discourses, and practices, whether political or faith-based. Abortion discourse still dominates media and politics in Ireland.

In the US, the divisions between a sequestered science and everything else on the curriculum go back a long time, not surprisingly coinciding with the rise of 'modernist' science as an institution in the nineteenth century. Gieryn (1983), in his classic paper on boundaries of STS, describes how early pre-Civil War holistic and spiritual explanations that were part of popular wisdom, begun to be driven back by a concerted educational effort to introduce what has since been known as 'antebellum science'. As with our bio-oratory notion, there was also increased explanation of scientific phenomena in public spaces as a means of popularisation with the hope of the eventual 'cleansing' of the 'superstitious old ways' in favour of an egalitarian worldview. However, today, junior high and high schools across the US enjoy a high degree of autonomy in a federalised education system. This is why jurisdictions such as Arkansas spring up in news stories when Intelligent Design (ID) is placed on the biology curriculum alongside evolution. To most educators, this is a return to pre-Civil War superstition. However, there is a strong argument that ID is less 'dangerous' to society than eugenics. A core textbook on the US curriculum was George Hunter's *A Civic Biology,* which contains no mechanisms for critique. Eugenics thinking was mainstream biology then, and was the 'moral sociology' of this text, influenced by Francis Galton. Eugenics was 'the science of being well born' (Hunter, 1914: 261). It is lost on our current standards of moral education that this was the textbook shelved in favour of creationism during the Scopes affair. We will come back to this battleground of ID in Chapter 5, looking at one set of schools in particular, in Dover, Pennsylvania.

It is well documented that the first textbooks that can be called 'biology' were Herbert Spencer's *The Principles of Biology* and T. H. Huxley's *Course of Practical Instruction in Elementary Biology* (Huxley and Martin, 1875), originating from Britain but translated across Europe. As well as their disciplinary and curricular development, these texts made a break from anatomy and botany, the twin preoccupations of previous biological education (Mansell, 1984). The texts, in line with the times (in much of Europe and the US), supported a socially inclined biology that 'treated' the undesirables of public feeble-mindedness and madness (Selden, 1991).

After the Second World War the global eugenics movement was abruptly ended. There was a deep feeling to separate *Weltanschauung* from life sciences, as though science was somehow made impure by politics. A German textbook of 1942 reads: 'The battle for existence is hard and unforgiving, but is the only way to maintain life. This struggle eliminates everything that is unfit for life, and selects everything that is able to survive' (Harm and Wiehle, 1942: 168). Given the historical spectres of eugenics and the recognition of a moral dimension to biology was recognised by the United Nations, any right to biological race distinctions in policy was removed (UNESCO, 1945). There is now a twenty-first century upgrade: Articles 20 and 21 of the Universal Declaration on the Human Genome and Human Rights (UNESCO, 1997) was adopted unanimously by

186 member countries of UNESCO on 11 November 1997. They outlined the ambitions for an education policy that recognised ethics in biology 'through the promotion of education in bioethics, at all levels, in particular for those responsible for science policies' (UNESCO, 1997: e1) while also 'raising the awareness of society and all of its members of their responsibilities regarding the fundamental issues relating to the defence of human dignity ....raised by research in biology.... ensuring the free expression of various socio-cultural, religious and philosophical opinions' (ibid. e1).

While the Final Solution is an unthinkable historical outcome of eugenics, we must guard against the exclusionary, while accepting that a separation of bioscience from politics cannot be attained. If we trace back to what concepts remain even of Nazi Germany, the Uexküll interaction of organism and environment remains in biology curriculum, as does *Lebensraum* – place – which was seen as critical to Nazi ideologies. Through a contemporary STS lens, biological artefacts and geographic space interactions blur the distinctions between 'the born and the made' (Thrift, 2005).

However, a reproducing curriculum fails to control narratives of biopolitics. It offers no extra power against the resistance. Adshead and Millar (2003) are of the opinion that the welfare state in Ireland developed post-independence as a direct result of what they call Catholic corporatism. Health, education and spirituality were exclusively under the control of the Catholic Church. This in turn, they maintain, impacted on public policy, and to this day there is no tradition of public policy oversight. The first victim of Ireland's economic crash was the Irish Council of Bioethics, discontinued in 2009. Is the answer then an education that attends to life political debate at times, at all grades in schools? Is this reflexive modernisation directly applied to education?

Outreaching for the body

As Mode-2 technoscience shapes our lives, the institutional dissemination that arises from it, 'Education and Outreach', has become highly strategised across the world. There are nanotech road shows, science centres and museums, open laboratories, fame labs, TV and digital media objects, plays, citizen science initiatives. In ways, these both compete with and contribute to contemporary bio-orators who add to this growing system of bioscience communication with David Attenborough, Robert Winston and Brian Cox on TV. Cox and Winston are both part of the UK science-communication-as-literacy movement, as are Caged Monkeys, often using a tweet-driven push towards a 'Science is Fun' narrative. As stated earlier in the chapter, they are a different type of public intellectual, a different breed of bio-orator than the speakers of the past. Education and Outreach is the latest professionalisation of science communication, artisans that circle the artists in the new engagement with STI.

The objections that can be made against the outreach concept as it exists are pedagogical and philosophical. On the pedagogical level, the use of a 'deficit model' type of dissemination of knowledge from institution to the 'unenlightened' does not fit easily into current PEST models (See Chapters 1 and 3). Since there are on-going programmes of outreach that are set to be funded, we must take care that they expand rather than narrow educational possibilities.

The conceptual issue of outreach, its tradition and colonialism, fits with a linear, Victorian natural history, which are similar to the critiques of public engagement by Fuller (2011) and Latour (2004a), already mentioned. Biology as a subject is still laid out in sections of a textbook, and has changed little over a century, where biology was 'the synthesis of zoology and botany' (Tarde, 2000 [1899]: 16). (There is at least an excuse for the Natural History museum in Dublin, a museum piece in itself, laid out and remaining in Linnaean system as it was in the Victorian age, when the museum was built.) Where the more non-formal engagement of outreach revolves around the 'text' as it were, then there is the potential for elitism. The outsider, the academically less-gifted, can feel isolated, as happened at times during my own school visits. Teachers in two schools that contributed to this study did everything they could to get their student into a programme associated with DCU. It was commendable how much teachers were willing to show me excellent work from other domains of science, However, to an outsider, there seemed to be a focus on those more academically gifted students.

In Ireland, the development of Centres for Science, Engineering Technologies (CSETs), Discover Science and Technology and the Programme for Research in Third Level Institutions (PRTLI programme), were part of third level research centres in direct connection with second level, part of large-scale funding that began with the White Paper (1996), and have been seen as the 'Irish turn to science'. Has their history ensured then that they are promotional rather than educational? Can PRTLI do 'ELSI' or is the higher and secondary education sector merely responding to criticisms from Craig Barret of Intel and others about the scientific literacy of Irish young, based on OECD and PISA (2012) data figures? We saw glimpses of the historical context for science education in Ireland earlier in the chapter when discussing links with Protestant Ascendency and the influence of Free State, Pearsean republicanism and religion in early state formation.

However, as Europe heads into the research era of Horizon 2020, it has placed great importance on science centres for public outreach. In Dublin's Science Gallery, part of Trinity College Dublin SFI/PRTLI, connections between science policy and concept of 'outreach' will need to share ideas across European centres, where there are different senses of the civic and public – a broader education that is a civic-minded, public pedagogy. What is also telling about how outreach might 'do life politics' is the line crossed between pedagogy and art, where embodiment is constantly sought, to make relevant, proximal and engaging. As discussed more in Chapter 4, often topics about the body become the sites of greater engagement. It is the pure relationship, the authenticity and self-actualisation recognised by Giddens in his defining of late modernity and institutions.

Despite this PR-enhanced outreach, however, there is one initiative that stands head and shoulders over any other in terms of discursive engagement – Debating Science Issues, a Science Foundation Ireland funded initiative, going strong at the time of writing. This is an intra-school debating competition about emerging technologies such as GM, nanotechnology, climate remediation and stem cells. Also, the two NICB researchers who ventured out into schools during my PhD research fulfilled an 'outreach' mission, of a kind. Industry and higher education institutes (HEIs) have a role to fulfil in supporting teachers, offering resources, packs and life-planning choices in terms of jobs, the official science version of what happens in a lab. Engineers (bio or STEM) can be part of the bio-orator tradition. But we must be careful about how much there is industrial and commercial control on education.

Since Darwin, *homo sapiens* is further displaced; we seem to have been knowingly trapped in our biology. But modern biotechnology, and genetic modification in particular, has given us power to change and control our biologies, our destinies. How can we teach that to our young? For such an important topic, should young people not have a voice in these new future branches in a strange, changing evolutionary tree (potentially evolutionary)?

Ultimately the bio-orators are needed, contentious though their views may be, but they are far from the full picture of bio-discourse. Theirs is the dissemination part, the one voice not always looking for a response. The progressive onward march of ideas about individuals and agencies, leading to how we perceive ourselves as human and hereditary beings against Nature, with unitary genetic markers, has had its troubled history. Even strategic funding favours a new age of biology linked to generational, aging and global 'challenges', relevance and the individual body remerges. However, with the exception of Attenborough's postmodern epilogue at the end of his shows demonstrating how Nature was captured by the cameras, why is the communication, embedded societal and discourse of biology rarely reflexively acknowledged in the educational field?

Smartly dressed, nervously anticipating though they may be, some of the more prominent guests in the Exam Hall at the key ESOF 2012 event awaiting J. Craig Venter – politicians, policy-makers and leading Principal Investigator scientists – are uncomfortable with each other. This was reflected in the barbed comment from Prof. Drury in his opening address, which indicated the wrong road turned in Irish policy on privileging applied science over basic science. Truthfully, where industry dictates policy, education is not always best served. Be that as it may, what is clear is that in Ireland, as elsewhere, the strategic common vision idea of biosciences for human betterment runs adrift with local practices and politics. There is the constant question of ownership of knowledge and what attitudes matter to the governance of bioscience in any case, and how are such things measured and used for biology literacy. It is to these issues we shall turn in the next chapter.

Note

1 Despite the globalisation of S&T as macro-structures above nation-states – as demonstrated by the Clinton/Blair HGP announcement – Ireland, in line with other countries, may still frame science for localised political gain (Jasanoff, 2005). This is discussed further later in the chapter.

Chapter 3

Biopinion

In the previous chapter, we tracked some of the assumptions, thought collectives, channels of knowledge and biopolitical practices that shape ideas of nation-state, and from these, the strategic agendas that influence how we teach our young people about reproductive and genetic technologies. This chapter looks at the role(s) of expert and public opinion, particularly public opinion, and the implications for RGT practice and policy. While opinion formation is an important aspect of argumentation, it is often given a dangerous tag in the physical sciences, as in 'mere opinion' when set against 'objective science'. Also, in politics, it cannot be underestimated how much a temperature gauge of public views is constructed, manipulated, managed, if not always deviously so. When we tie opinion to literacy, there is a certain input/output simplicity assumed that does not exist in the real worlds of education, or indeed political action. Opinion on bio-technical topics cannot be formed, or informed, by biological literacy alone. We look in this section of the book at what it might mean to 'track opinion' from the school room all the way to some type of political governance, of social order and parliament. We cannot easily track an idea – each locale of information exchange is highly contextualised on such an imaginary pathway from school to governance. What we do here instead is analyse the sites distinctly and trace patterns through each site, particularly where opinion-construction appears to be happening. There is no fixed public knowledge or fixed opinion. The analysis in the previous chapter was highly cognizant of the strong link between large 'P' Political will and perceived public opinion. The role of science is important as historically facts and action are legitimated by its power and intertwined with political action, and therefore some kind of political/public opinion axis (Jasanoff, 2005; Shapin and Schaffer, 1985). Bioscientists with vested interests in the agricultural or health sectors may seek public trust and increased bio-literacy. Mainstream politicians tend to want a quiet life in matters life political. The volatility of public response on genetic technologies in US and Europe ensures this peace is shattered (Bauer and Bonfadelli, 2002; Hornig Priest, 2001).

By the end of this chapter, we will have explored life politics in more detail through the key concepts of *witnessing* and *mediation*, but first we need to look at the meaning and constructs of public opinion, often offered, and deferred to, in political representation and its frequent use in biotech discourses, as well as how these constructs are used. The changing nature of science and its traditional counterpoint to opinion is then analysed, given the multiple-perspective issues that face RGTs as well as new conceptual approaches in STS, materiality-focused political theory and biopolitical philosophy.

Medium-to-large-scale public opinion is often gathered on everything from genetically-modified crops to stem cell research, through the work of George Gaskell and Martin Bauer and others and their contributions to the EC Eurobarometer reports, as well as studies of attitudes of young people to the biosciences (Bauer and Gaskell, 2002a, 2002b; Gaskell *et al.*, 2000; Gaskell *et al.*, 2002; Gaskell, 2003; Hornig Priest, 2001; Jasanoff, 2005; Torgersen *et al.*, 2002). These representations of publics, as well as the mechanisms of expert opinion representation for biopolitical knowledge construction, are part of a growing area of interest in what I call here 'biopinions'. However, we will hope to see nuances here between attitudes, literacy and opinion as we analyse these sites of inquiry. The crucial element is how these pathways, with all this decision-making complexity, contexts and 'muddle' (Ravetz, 2005), can have a place in curriculum science.

What is 'opinion'? Why does it matter (to science)?

On 16th June 2003, Jamie Whitaker's birth provided a lifeline to his brother Charlie. Charlie had Diamond-Blackfan anaemia, a rare genetic disorder requiring him to undergo a blood transfusion every three weeks. The boys' parents, from Sheffield, England had to travel to the Reproductive Genetics Unit in Chicago to undergo IVF treatment with pre-implantation genetic diagnosis (PGD) to select a male embryo with a genetic match for Charlie in a last bid for treatment. Jamie's umbilical cord would be part of that treatment. The Whitakers had pleaded their case with the Human Fertilisation and Embryology Authority (HFEA) but were not granted treatment in Britain because, under existing guidelines at the time, PGD was allowed in rare cases where the conceived child was at risk of disease. Since then, guidelines have changed to allow tissue matching in exceptional circumstances. However, not before one case was brought to court on two occasions. The Hashmi family, in similar circumstances to the Whitakers, requested PGD for a so-called 'saviour sibling' to their sick son, Zain. In this first test case, the HFEA agreed to allow the Hashmis to carry out the procedure for a baby son. Thus drew hostile media and interest group reaction. The Whitakers subsequently were refused similar treatment. Many commentators blamed the HFEA for bowing to 'the court of public opinion' – all the more of an issue now in the age of Twitter – but the decision in both cases appears to have been based on a more conservative genetic screening procedure (Sheldon and Wilkinson, 2003). For the Hashmis, the condition was hereditary. The justification then was that screening was in the obvious interest of the future child who would develop. For the Whitakers, it was not a hereditary condition, and no screening was required as a preventative measure for the prospective child. This was a technicality that made the difference, although many claimed in media commentary that the controversy following Hashmi was also a factor. However, the House of Lords had overturned a previous decision made in favour of the claimant in Comment on Reproductive Ethics (CORE) v HFEA which challenged a HFEA decision to allow treatment for the Whitakers. The trial and subsequent appeal centred around fears about what was meant by a 'suitable' embryo and the prospect of 'social sex selection' (Human Fertilisation and Embryology Authority (HFEA) Ethics and Law Committee, 2003). CORE also asked the courts and the public – it was hotly debated in the media (Challand, 2003; Kenny, 2002) – to consider the potential identity question for James and others like him: was he born for the purpose of saving Charlie?[1]

Unfortunately for the Hashmis, frustrated after several unsuccessful IVF attempts, they gave up. They were the first couple granted the HFEA license for sex selection. The UK HFE Act was subsequently passed in 2008 bringing more clarity for couples, including same-sex couples, on the issue of sex selection of embryos. The Whitakers now have three healthy children including Charlie, who was given the all-clear in 2005. At the time of writing, Charlie is a healthy fifteen year old, while Jamie is ten. In the same year that Charlie's saviour brother arrived and of the Law Lords' decision that PGD in the Hashmi and Whitaker cases were lawful after all, the *Irish Times* education section reported a 'controversy' among teachers about the appearance of an Irish Leaving Certificate biology examination question which asked students to propose cases for genetic testing after IVF (Downes, 2005). The procedure, for any condition, was quite a new concept to Ireland at the time.

These accounts demonstrate how recent RGTs have advanced at a pace barely matched by the legal, social, cultural or indeed educational arenas, and differing national attitudes. The issues thrown up by these technologies have changed relationships over a relatively short period of time making regulation difficult and creating social and ethical complications. There are new types of relationships emerging, new identities, summoned up by media terms like 'designer baby' or 'saviour sibling'. Young people are growing up in a world complicated, perhaps enriched, by bioethical dilemmas associated with technology. As anthropologist Marilyn Strathern (1996) says, there are new kinds of social relationships being created that are tied up with technology and forms of commodity that change family ties and kinship. However, these new relationships resulting from RGT extend across all webs of society, not just the family.

In the domain of patent law, US Supreme Court case *Diamond v Chakrabarty* is the seminal case (Jasanoff, 2005, citing *Diamond v Chakrabarty*, 1980). Living material suddenly became constructed as a patentable object in a landmark ruling in 1980. The *pseudomonas* bacterium was developed by Indian-American molecular biologist Ananda Chakrabarty using a plasmid vector to create a rapid oil-metabolising species, useful for slick clean-ups.

In commerce, genetically-modified (GM) food products are regulated by the Food and Drug Administration in US and the European Food Safety Authority in the EU. Among the first products on the market in the 1990s were GM milk, tomatoes, and maize (Hornig Priest, 2001). Since Berkeley researchers first developed 'ice-minus' *Pseudomonas* using recombinant DNA techniques that could make plants more resistant to frost (Jasanoff, 2005: 98), the movement from university or academic origins to commerce is a messy one; indeed a reflexive modernisation or Mode-2 science view of science would present the case that its existence cannot be other than 'messy' (Nowotny *et al.*, 2001; Ravetz, 2006). The Berkeley researchers would not fit easily into our 'bio-orators' list either from the previous chapter – they failed to convince NGOs that they were anything more than corporate shills. Cue many trials disrupted by physically uprooting potatoes, a political, performed act that speaks louder than words, also used in similar ways, and with similar GM crops in UK, France and Ireland. Commercial interests are still bound by local regulations, whether that is in the EU, Asia or US law. Politicians and policymakers realise quickly how crucial public opinion becomes once politicisation – and mediation – of an issue is increased. Whatever about the symbolic level of environmental, health and taste (Eder, 1996; Murphy, 2010), once the issue becomes a consumer health one for food, the whole game changes. The volatility of public opinion will be sharply felt. Suddenly the interest groups are mobilised and public trust in multi-nationals and commercial ventures in the area – the habitat of innovation and 'translation' of science into product – is eroded. Bioscience issues are very public issues at every stage.

The new genetics is a complex range of areas, with apparent cultural resistance globally. We cannot say oppositions to stem cell research, the use of biotechnology for climate change remediation and GM foods raise the same issues, but there are patterns of similarities across many resistances, what is often referred to as socio-technical issues, involving deep-held notions of Progress, Nature, Science and God. GM food has had a particular large amount of STS research conducted on variants of 'attitudes' for want of a better term (Einsiedel, 2001; Hornig Priest, 2001; Jasanoff, 2005; #gmpotato Community of Inquiry, 2012).

However, there is more to the integration of biology as a concept into public/population minds and practices than attitudes. Biology has never been such a part of culture as it is now – the pervasiveness of the double-helix in iconography (Nelkin and Lindee, 2004), the use of biological terms in common language such as 'nucleus', 'cells', 'in the genes'. There are two issues with the use of opinion with regard to RGTs and all biosciences: 1) how epistemologically it deals with 'Science'; and 2) how methodologically and conceptually public opinion in particular is collected or engineered.

Let us deal with the second issue first, being capital 'P' Politics, and we will then ease into science and opinion. The meaning and construction of public opinion is a common method of informing political representation (for the initial process of representative speaking anyway, whatever about the form and direction of politician's intentional discourses afterwards – I will look at Pitkin (1967) and Disch (2008, 2010) later in this regard). It is an interesting question why it has been brought into biotech discourses so frequently, and how it is used. We have long known that polling is a means of constructing publics to bring into play an act that rhetorically 'gauges' a feeling 'out there' with the view to influencing political and media action and instigate media agenda-setting (Lippmann, 2009) [1921]). Surveying can be a crude instrument however. The trouble with surveys or polling for cultural nuances for genetic technologies is well described in Hornig Priest (2001), which includes the restrictions of sampling and the accuracy of elicited responses. We must give great attention to whose opinion we are looking for now that we can also use online and social media methods with increasing numbers. What wisdom of what crowds are we talking about when the internet has so much misinformation and *dis*information? Yet the political reality is that opinion may be gathered and be accepted without question, strategically constructed in surveys and newspaper polls. The truism is often used: 'When my opinion changes, I change the facts'. Any undergraduate student learning the techniques for research methods realises quickly the nuances of question order, self-reporting and the presence of downright fibs. There are sophisticated ways of countering these on a questionnaire or by revealing of questions on an omnibus telephone survey, but when it comes to the heart of it, we cannot know inner workings of individual identities.

The trick is not to try. The front stage (Goffman, (1990)[1959] offered by respondents has the same value as a voter in an election: it is the right thing to do in a democracy, they have been called on, they are part of an appellation and an answer is required and a resulting configurative construction may mean a macro issue, or political structure, shifts. The responses work exactly as the term 'Eurobarometer' describes it, broad indicators of continents of opinion. They also work much like the students' rhetoric of opinion on Chapters 6 and 7. These are not necessarily, or at all, beliefs. Even a future participatory method of e-polling of the masses will have its difficulties when we analyse what the votes may mean, let alone what they would say to political leaders. However, if we need to consider life politics in a future Parliament of Things, political representation of 'outsider' propositions is key.

Opinion can be better analysed around action as a co-construction of knowledge, active communities, not easily analysed without nuance, shored up, simplified and parsed for easy access for journalistic dissemination or for data-mining. The theme of distrust mixed with misinformation in the Eurobarometer reports reflects just part of what 'attitudes' actually are. An EC Science Research Development report on cultural and social attitudes of the young (European Commission, 1997) points out that viewpoints expressed on biotechnology may sometimes be tied in with accepted norms, the answers the respondents think are expected to give, and poses the question: where does biotechnology fit into current political and social ideas like communitarianism and liberalism? Studies on the nature of attitudes (Evans and Durant, 1995) show that people make distinctions between different areas of science when reflecting on whether it is worthwhile or morally justified. This is not just a European phenomenon. However, consensus models derived from Scandinavia have demonstrated to be effective and 'travel-friendly' across continents (Einsiedel, 2001) when it comes to GM technologies. Something more nuanced and sophisticated than digging for biopinion is needed. The Eurobarometers, while useful testing large-scale 'pressure' units out there in the atmosphere, can have a tendency to ignore the 'temperatures' of contextual cultural meaning-making and lived experiences. I have argued elsewhere about the Latour notion of 'no controversy, no issue': that communities of concern need to build around a socio-technical issue in order for deep non-expert, public involvement (Murphy, 2010). However, the idea of an observatory of opinions is a powerful one, particularly in the context of what European research programs are now calling Future and Emerging Technologies (FET) such as the nano-bio-info-cogno convergence technologies mentioned in Chapter 1, where there are great uncertainties. Martin Bauer and George Gaskell and their colleagues are well aware of this and are involved in the tracking of what one might call 'co-product opinions'. Interactions can be mapped out in sites of practices between a notional 'science' (which is in reality, a multitude of economies, thought collectives and practices) and 'society' (to which our publicly elected officials and public servants must, in theory, be accountable). Opinion then should be set within the context of these multiple literacies.

Looking at various angles on the issue of assumptions about science's apparent rejection of 'mere opinion', we see, first of all, that contemporary science communication theories and practice attempt to break down the barriers between scientific experts and various types of public. STS has been influential here, linking up streams of excellent research from the fields of the history, philosophy and sociology of science to demonstrate contexts, co-production, the prevalence of materiality and the sometime muddle of science. After the Enlightenment, the default setting for scientists was to separate epistemologically scientific facts from everything else. Mode-2 science seriously challenges this. Publics are problematic for institutional science in many ways. Harvard STS Professor Sheila Jasanoff (2005) calls it *civic epistemology*, how political systems become shaped around knowledge pathways of technoscience by experts and non-expert publics, the cultural milieu for knowledge production, if you will. This is not a world away from Warwick STS professor Steve Fuller's *social epistemology*, the social foundations of knowledge, differing only perhaps in how many obscure and contra-scientific views Fuller allows into his scientific legitimacy (Fuller and Collier, 2008). In the context of our human genetics and futures debate, Fuller (2010; 2011) is one of the main critics of the public engagement agenda – non-expert opinion need not be changed, Fuller decrees, and in his view this is the objective, implicit or otherwise, of these so-called Public Engagement with Science and Technology (PEST) initiatives. Certainly, the widespread public mistrust of the *GM Nation?* public engagement exercise in the UK testifies to an indifference, if not outright hostility, to the motivations of the engagement strategists for the development of genetically-modified foods, as participants suspected New Labour social marketing (Gaskell, 2003). Focus groups received bad press for a while in the UK. Also we see 'deficit model' assumptions in the dissemination bias of STEM programmes (Trench, 2008).

For Steve Fuller, it is up to an evolved model of Enlightenment science to take on board social epistemology (Fuller, 2010, 2011). Fuller, to be fair, makes these comments in the context of convergence technology and human enhancement, and the tendency for strategic Big Science to try and persuade people of its virtue. To chime with our analysis of the previous chapter, it is, after all, a type of collective bio-oratory in its own right, necessary cycles of competitive PR in media spheres and brought to schools. Much of the scepticism of public engagement initiatives from Durodié (2003) and others centres on how core-set technological expertise loses its value as it moves outside through efforts of explanation or by popular science. In other words, *The Mismeasure of Man* (Gould, 1996 [1981]) might tell us much about misapplied policy leading to bad science on race, but it is not real science. Such a view would also argue that Angrist's (2010) *Here Is a Human Being* is a good summary for the uninitiated on genomics, with some humour and philosophical questions, but hardly real biological science. The richness and depth of data is watered down by popularisation. Opinion then appears on the edges of, and outside, deductive reasoning and scientific rationality. However, Hilgartner (1990, 2000), and many STS and science communication scholars since have demonstrated quite categorically how the construction of viable, verifiable, and even inspiring truths can occur when the half-data of an HPLC or PCR process in the lab meets laboratory rhetoric, which in turn meets curiosity and comment from outside the lab. It is not even a pebble-in-a-pool type of outward communicative motion. Biology communication and education of RGTs not only are enhanced, but *need,* I would argue, film, media debate and the real family issues of RGTs, from the grant proposals for new RGT techniques, to the media coverage of a breakthrough to the angry interview with a member of Genetic Concern. Latour's things and propositions for Parliament of Things are created with these tools. (And let us not forget that our progenitor in this section on opinion, Lippman (1993) [1921], used his 'phantom public' model as an argument for keeping the barbarians at the gates. Latour, a champion for Lippman's phantom public, has not addressed this elitism.)

So although Walter Lippman strongly warned the opposite quite a while ago (Lippman, 1993[1927]; 2009[1921]), the idea of finding out public opinion on anything was once seen as a straightforward task. Learning 'what the public thought' would mean identifying motivations and indicators to enact behaviour change – all expected to have a positive influence by then increasing the knowledge input 'at source', leading to greater scientific literacy. (In line with other US pragmatists, his contemporary John Dewey, a formative science education proponent saw the public as 'the emergent effect of a problem' (Bennet, 2010: xviii).) Current science communication writing constantly refers to this as the 'deficit model', and uses studies such as Evans and Durant (1995) to demonstrate that more information can directly lead to more public questions, as indeed one would expect from sub-politically engaged individuals or collectives. And here is where things get hairy for the positivist, 'Modern' bioscientist. Any amount of STS scholarly work will tell us biotechnology exists as much on the symbolic realm as it does in the lab (Bauer and Bonfadelli, 2002; Eder, 1996; Erickson, 2005; Fleck, (1979) [1935]; Nelkin and Lindee, 1995; Wynne, 1992). The 'thing' (that Latour word again) that publics need to find out about is different in each locale, and it evolves. Mediation, by traditional broadcast and print media as well as social media, and the discursive practices that take place across different genres (Fairclough, 2003), leads to an agreed RGT or biotech lab procedure. In other words, the BRCA1 gene will mean a different thing when written up in a Horizon 2020 grant for R&D predictive medicine funding than it will in the boardrooms of Myriad Genetics, or when argued in the US Supreme Court as part of patent proceedings, or indeed when discussed at a press conference of the world's journalists by the actor Angelina Jolie just after her double mastectomy to prevent future breast cancer. This latter event alone brought embodiment, life political planning and symbolic engagement to an issue that no amount of PR or patent documentation ever could. For many women, it has enhanced their reality. And the bioscientists of Myriad, which holds the BRCA1 patent, will have more work to do. Opinion can never truly be separated from the processes of science. And this is without speaking of opinion-forming in the lab, where a mixture of translations take place from instrument to database that involve colloquial, lab-based esoteric language mixed with the exoteric of 'the outside world' (Erickson, 2005; Fuller; 1997; Fleck, (1979) [1935]; Latour and Woolgar, 1979).

Often it is not until there is a sociotechnical crisis that the idea of a 'public' emerges. As a Demos report states – perhaps over-stating the case, but only by a little – there is no opinion formation until there is a summoning, an Althusserian interpellation, if you will (Wilsdon and Willis, 2004). In other words, no one has an opinion until asked, and the whole composition becomes co-created. The Mode-2/post-normal thesis (Nowotny *et al.*, 2001; Ravetz, 2005), which we follow here, expects structurational changes (Giddens, 1994) from technologist to consumer and back again in this co-production. The fact that public opinion is often still called on as part of policy-media agenda-setting, by surveys, interviews and focus groups, suggest that a Mode-1, deficit model idea of science as distinct from society still exists. However, even if we were to consider the Mode-2 version, once we move out of core-sets of expertise (Collins and Evans, 2002), we still would not naturally expect high levels of the type of biology knowledge found in textbooks and science websites.

This interpellation, this summoning, is at the heart of deliberative communication between experts and non-experts for policy, the Public Engagement with Science and Technology (PEST) movement since the turn of the millennium (Irwin, 2006; Jackson *et al.*, 2005). It has been well argued in the science communication literature at this stage of the cultural, economic, democratic, participative, and indeed educational reasons why all publics need to be involved in the practices of science learning (Osborne, 2000; Stocklmayer, *et al.*, 2002), technoscience (Irwin, 2006; Haraway, 1991; Gibbons, 1999; Gibbons *et al.*, 1994; Nowotny *et al.*, 2001; Wynne, 1992) and megascience (Ravetz, 2005) happening all around them. When we see the history of the twentieth century alone, the erosion of trust in large-scale, disruptive, and, as it turned out, harmful sociotechnical projects – DDT, scientific experimentation on institutional inmates, nuclear incidents, and more recently genetic technologies – then there is justification in the argument that it is our obligation as part- or fully-public funded vested interests in science to not just disseminate, but in this recent paradigm of knowledge co-production, become *partners* with those with other forms of expertise and no expertise.

This brings us closer to analysing the meaning and constructs of public opinion when placed beside biotech discourses so frequently, and how it is used, how it – rhetorically, in swathes of practice – differs from bioscientific expert opinion, and perhaps areas where they are one and the same. In this battleground, epistemologies of reason should not preclude resistance to dominant scientific paradigms. If a family believes a disease is a curse from God because of wrong-doing, we scientists-communicators-educators need to tread carefully about how we deal with this. We should not always throw out what appears to be gibberish if performed in front of a live audience. DNA code terminology is also gibberish to most audiences. There has to be some context – different discourse sites require different genres and rule of engagement (Murphy, 2010). Using Latour's (2003) translation idea, like the classic sociolinguistics of Laclau and Mouffe (2001), we see that the bioscientist, the patient and the consumer and other actors negotiate their positions. It is the 'making equivalence of non-equivalent situations' (Disch, 2008: 95 quoting Latour, 2003) so that all or some of the connecting actors can use the (new) raw materials to translate or mediate further (see the 'front stage/back stage' descriptions of Beacon CARE Fertility in Chapter 4). In science education research, as we will see in Chapter 5, the STS movement seeks to include the use of values in science (Ratcliffe and Grace, 2003) and argumentation (Chowning *et al.*, 2012; Driver *et al.*, 2000; Jiménez-Aleixandre *et al.*, 2000; Osborne *et al.*, 2004; Parker, 1996; Sadler and Fowler, 2006; Simon *et al.*, 2006) as part of a wider scientific citizenship.

Outside of education, in wider society, there is varying interest in media stories and science exhibitions about origins of life, environmental science and biodiversity, forensics and any kind of monster creation. Content providers contain 'science' sections accessible on all types of handheld devices. It is where most people consume science, as interested readers, audiences and viewers. But the idea of pedagogy is unlikely to be found in such a concentrated way in these spaces, with the exception perhaps of matters specifically categorised as 'science'. Science news media informs rather than educates, entertains rather than spurs action. The tradition of science coverage has been one of informing or promoting (Hansen, 1994). The agenda-setting role of media on any topic has its place, as is well documented (McCombs and Shaw, 1972) where readers, viewers and listeners have their opinion shaped for them as it were, but we must not expect our media to respond sufficiently to long-term issues of genomics and environment (Hornig Priest, 2001). What we expect is short-term commentary and accurate reporting on a pressing sociotechnical issue, especially one with social, health or environmental risks (ibid.) Even in industry, trainers may train, but pedagogy is the domain of the teacher and academic scientist. The conceptual change – moving a child from preconception to mental reasoning – and the inspiration that the education system is supposed to provide often does not occur so easily within the rest of society. On 'the other side', within schools and institutions of education, as we saw in the previous chapter, the school and its curriculum becomes, through history and ideology, isolated, cut off, halted before it reaches the practice of civics.

Trench (2008) and Bultitude (2011) make strong cases for the use of the internet for informal/non-formal channels of science consumption, dialogue and debate, if not for public interactions with science journals given the prohibitively priced firewalls of academic publishers, a topic of growing concern for many in the science communication community. While studies in the early 2000s show that television had displaced newspapers as the main news information source in the US and Europe (Bauer and Gaskell, 2002a, 2002b), given the speed of change of young people's usage patterns for multiple portable devices, online games and TV programme streams, it would not be surprising to see that the internet has displaced them both. In a social science article for the *Science* journal, Brossard and Schefeule (2013) found that 60 per cent of respondents in their US survey depended on the internet for information on a range of topics, and 90 per cent of internet users went straight to search engines as their first port of call. Only 12 per cent got their science from traditional print and broadcast media, the study claims (ibid.). They also found that hostility to a particular emerging technology grew as the internet boards' trolling grew (ibid.). Accepting the weaknesses of a survey (pressure points, not beliefs), this still has huge implications for the future of science journalism, as well as the development of knowledge construction, were epistemologists to worry about the shallow waters of click-throughs. What it means, however, is that we need to pay particular attention to how easy clicks and screen swipes are but a starting point to opinion formation and argumentation within the *kakosmos* (Latour, 2004a) of immanently formed opinions on the internet. As science communication scholar Fellicity Mellor has said, in response to the Brossard and Schefeule findings, the challenge is for the strategic science community to expect and plan for the risk of public debate on new media, not try to control it (SciDev.Net, 2013). For citizens in all parts of the world, developed and developing, this issue is more about media literacy than science literacy (ibid.) Again, reflexive modernisation that incorporates life politics demands this level of engagement and confrontation (Giddens, 1991; Lash, 1994). The web then is *the* emerging forum for opinion. As analysed further in Chapter 6, these are part of a combination or resources of introductory opinion forming and knowledge-production (Gamson, 1992; Gamson and Modigliani, 1989).

How easy it is, however, for a young person on a smart phone to jump between topics. When young people are asked their views, perhaps based on this media, they 'position' themselves, and this positioning, I contend, is *immanent* (Davies and Harre, 1997), the act of the discourse itself, and how ambiguities can evolve even from multiple understandings of a single phrase. Immanent phrases are borrowed, repeated from communication practices elsewhere, other conversations, or media representations. The practice approach takes on board the discursive practices in presentation and meaning. In common with modern qualitative inquiry, practice recognises how the body is tied into discursive and non-discursive interaction, as Giddens' life politics also addresses. Schatzki refers to practices in their broadest sense as 'arrays of activity' (Schatzki, 2001: 2) or 'a nexus of (non-discursive and discursive) practices as body/knowledge/things-complexes' (Reckwitz, 2002: 258). We will see further in Chapter 4 how this performed nature of learning and learning environment, the *habitus* in locales, is vital to understanding responses to Big Science concepts.

But it is to the big -isms of our emancipatory past that science should not be afraid to turn, and that might include Marxism, gender, nationalism, race studies, environmentalism, and more contemporary movements like veganism and Fair Trade. These topics facilitate a way of educating ourselves and moving forward as part of 'colonising of the future' (Giddens, 1991). This is the taking control of our futures, and using it to controlling effect today. So we colonise with blood and family lines, and by extension, genes. Marxism has it, progressive republicanism has it, environmentalism has it. Unfortunately fascism also has it, and so, in an age that tries to hide ideology, we must guard against totalising ideas.

And political opinion is needed, in the strange entanglements and rhizomes of action, leading from the embryo on a petri dish ready for fertilisation to the reflections of a grown-up born to IVF to the press room in 2000 of the announcement of the Human Genome Project. Co-construction between experts and non-experts, through 'issue events' (classroom, courtroom, computer model, etc.) of matters of concern, rather than matters of fact (Latour, 2004a). The routine practice of the everyday becomes modified, perhaps subtly, perhaps disrupted. And for regulation of emerging technologies like embryo cloning or advanced genetic screening, who decides? This is at the heart of life politics. In the classroom, learning becomes political action, at street level and hedge school level. Can we ever have political systems that facilitate participatory democracies complementing political representation in *res publica*, that is, public things? (Latour, 2004b). Not knowing anything of our future as a sovereign in Europe and the world we must continue with new ways of looking at teaching, learning, a *life political pedagogy*.

To focus on what scholarly opinion from STS has to say about opinion, we see traction around issues of governmentality, the Foucauldian phrase for micro-control of bodies, to governance and the use of propositions and things in this governance. How do STS accounts view the forms of how opinion/fact distinctions are played out and constructed, from Boyle and Hobbes, from Schaffer and Shapin, Latour and the like? Social order and governance was the preoccupation of the elite thinkers of natural philosophies and public in the seventeenth century (Shapin and Schaffer, 1985). Public opinion as we've seen, as opposed to expert opinion, is constructed in particular formulaic ways for political analysts. How easy it is to strategise the total, by bringing in what 'the public thinks', what it 'believes'. Pedagogy, as Steve Fuller explains it (Fuller, 2011: 76), arises out of Cicero's rhetoric, the expression of *humanitas*, preceding government, systems of the good life and ethics to live by (see also Campbell, 2008: 27). But in traditional society of Roman times, the basic social unit was the *oikos*, household or family (where now RGTs extend and break traditional ties) as opposed to the *polis*, the city-state. This is similar to the Habermas version of pedagogy (I believe however we are beyond any singular notion of one public sphere; for Sloterdijk (2011), an oppositional kindred spirit, there are many). This pedagogy of the young has its roots in Freire (1989)[1970]; but the arguments of a complete dialogic learning are social practices, not subjectivities (Billig, 1987). There is an emptying out of the inner world.

Public opinion, perhaps social opinion, is needed by the *polis* for political representation for the initial process of representative speaking anyway, whatever about the form and direction of politicians' intentional discourses afterwards. In political theory, Lisa Disch (2008, 2010) has been struck by how Latour's STS of objects, instruments and data of the lab are relevant to mainstream politics. We are beginning to re-see what Hobbes and Boyle were arguing about. This is not just scientists as political spokespersons, as they must be for RGTs and climate change politics, but STS theories have moved toward object-orientation, preparation for how we might imagine a Parliament of Things, even if it is in a rhetorical sense, a representation of data and matter (ibid., Bennet, 2010). When we get into this territory of things with action, Fuller (2007), in a review of Latour's *Politics of Nature* accuses him of succumbing too easy to French political preoccupation with linking public to representation, while US politics is concerned more with accountability to the public from the body politic *contra* the politics of (future) bodies. The distinctions I make here, drawing on Pitkin (1967) as distilled through Disch (2008) are between democratic and Republican founding worldviews, and are key here, as we saw in Chapter 2 – Ireland is also a Republic, so post-colonial representative politics is enshrined in the Constitution. The main flaw in Latour, according to Fuller (2007), is seeing the same type of representation happen in science and politics. The technologisation of the process of politics, its search for a good 'common world' is in Fuller's view innately conservative.

What we have now is a crisis of political representation, but is it a crisis when you already know the representative class of politicians are not representing us through proxy, and never did (Hanna Pitkin (1967) has been saying this since the 1960s). A way forward, which will unfold with later chapters, is a further 'socialised' contract (Hobbes, Rousseau, Shapin and Schaffer, 1985) as Gibbons (1999) argues but one that brings *bíos* into *polis*, a recognition of our Nature and Culture as inseparable, and above all, for an effective pedagogy of the internet generations, a removal of presentism, and involvement of those colonies of the future that Giddens speaks of. Could we make this present-future link formal and binding? This holds for enhanced humans as it does for resilient environments, where things and objects 'stand for' (Pitkin, 1967: 1) themselves in the future assembly, 'acting for' rather than mere representing, as conventional parochial politics currently demands in a republic. 'Opinion' plays a big part in how governments, policymakers and technical elites see public response, and there are implications for who controls what. The speech act, as per Latour (2004b) on Whitehead, is 'where the many become one'.

The question then is: can the Leviathan (Hobbes, 1651; Shapin and Schaffer, 1985) include social movements, advocacy groups and non-government organisations (collectively known as 'Civil Society Organisations' in the European Research Area)? Social movement and media research demonstrates that these groups, when pitted against a specific sociotechnical issue like GMOs or incineration, mobilise actors in the raw politics of prevention, lobbying and public awareness-raising (Bauer and Bonfadelli, 2002; Bauer and Gaskell, 2002a, 2002b; Gamson and Modigliani, 1989; Hornig Priest, 2001; Jasanoff, 2005). Besides the life political specifics of the sociotechnical issue itself, there is the added energy of being in a minority where no one appears to be listening, what Noelle-Neumann (1993) called the 'spiral of silence'.

It was clear in Chapter 2 how mobilisation of collective opinion and action depends on key individuals within nodes, which are in turn, within networks of biosciences. Those not in earshot, as it were, not there to witness, or to be taken in by the thought collective (Fleck, (1979) [1935]), will form their own social movements, if the sociotechnical issue is 'hot' enough for such public mobilisation. So of all the discussion about fiction in the chapter following this, and its importance to life politics pedagogy, none is as little recognised as the one that states that there are a multitude of opinions 'out there' that need to be harvested (from adults to inform policy and prevent political crisis) or realigned (in the case of pre-constructivist educational models for children and the deficit model of science communication for everybody else). We then begin to see a connecting of two locales – students become politically aware, that is life political, while contributing, again politically, to the exchange of workable and relevant opinion-forming on those very life processes and *technia* that are most political.

Life politics is the new politics, but not really new. It is that which was 'always already' there. It turns out the unsavoury upstarts at the gate have broken through and are welcome after all. The advice from either government or scientific advice along the chain could not have helped the Hashmis and the Whitakers. They needed to have some public visibility, making things public, helping the enactment of fleeting public opinion. And this involved courtrooms and media appearances for those caught into the compositions, which were also strewn with life politics. Life politics is always future orientated. We will see much later in this book something about giving and taking for the future. We will also see how Thompson, drawing on Rawls' idea of 'just savings' for future individuals/societies, offers a secular notion of giving of the future with some receiving in return.

Argumentation and opinion: political and philosophical theories for biology pedagogy

So let us take stock – the political of biology may not be so obvious until we position biosciences and life up against the macro of Big Science linked with policy, and the micro of person relationships with body, organisms, genes and molecules. But all of this is the first step for the *philosophical* questions entering into the biology classroom, but which also demonstrably is politically entangled. The philosophical questions are crucial, and students can be encouraged into the biological education sphere quite easily.

The obvious starting points are Darwinian theories, the separation of humans from animals and the natural world, and the new thinking of sociology from cell level up to macro level ontologies just described. If we consider the emancipatory theories on which life politics is founded, the struggles in theory and politics worth ideas of biological determinism becomes the most important philosophical idea in which to raise critical issues for children and young adults. The negative applications of Social Darwinism, from Galton to Herbert to Huxley, as outlined in Chapter 2, would be a point of departure for classroom discussions, as well as other past and potential future species of human (Fuller, 2011). The control that was required of Social Darwinists was a technology in its broadest meaning, a striving forth of the will of the elite and impacting on humanity in society in its applications to race, gender, sex and sexuality.

In these pages we have referred back to an earlier historical event for science, Boyle's Air-Pump. Science, to give it its impossible totality, is messy, uncertain and removed in its objectivity. The quadrat, the petri dish, the chromatograph following PCR for genetic analysis, supposed to be removed from human contamination. Shapin and Schaffer (1985) were among the first to show how science and policy separated from each other in the seventeenth century in *Leviathan and the Air-pump.*

Future scenario projects map forward (Groves, 2006; Guston and Sarewitz, 2002) set against the classic STS of multiple potential pathways of translation (Callon and Latour, 1981) demonstrate constant splitting where the most eloquent theory wins out. We cannot go back to every split for biology class, but besides rival Darwinian theories *in his time* – as teachers should constantly stress! – the other great split surely is rationalism and empiricism, as embodied in Hobbes' ideas and Hooke's Air-pump, and in many ways this will also deals with Fuller's Weissman split, the reintroduction of values, social purpose and human meaning to biology content.

What is common to all our tasks for science students is for student biologists, out in the fields for ecology, or at the lab bench, or on their laptops to bring a sense of wonder to their work. Even if is not *their* wonder, there is a history of Romanticism of (either) Rousseau, Thoreau, Whitman to the field; ideas, abstraction based on what they see, but we can see so much beyond the object now. This is to be a true witness, an authentic witness, following Giddens (1991), a creditable, credible witness (Shapin and Schaffer, 1985) if they are to be reinforced with theories of the good. It is the theory and order of the good in society, and linkages in nature that Hobbes would have evolve. The old idea, Humboldt's Nature as Morality, used the word eloquence, which Latour (2010) invokes for a theory whose time and composition is in place.

The great debate on evolutionary biology is captured for pedagogy neatly among STS and science luminaries in *Alas Poor Darwin*. Philosopher Mary Midgley (2001), in her chapter, explains how meme came from mimeme, a model of imitation, which in turn came from phoneme, a unit of sound developed by linguists early in the twentieth century. The hope for unification of the sciences was first influenced by how the work of Newton in the sciences and Locke in psychology could be combined, but was superseded by the rise of the physical sciences. Midgely is uncomfortable with the notion of memes existing, like Plato's ideas, independently of human thought or action. She claims that the Dawkinsian descriptions of memes are phenotypes because they are 'things in themselves', even if they are constantly changing and evolving, and if you approach this from systems theory, a meme is possibly less a 'thing in itself' than the intricate biochemical nucleotide pathways and processes that we call a gene.

Implicit in these bifurcated debates (Whitehead, 1920) of nature versus nurture are other dichotomies, such as Heidegger's classic *humanitas/animalitas* (Campbell, 2011: 28), how we are human and animal, but through the genetic codes and theoretical constructs now taken into bioinformatics and memes that run through *Alas Poor Darwin* is also the distinction and tension between *humanitas* and *informare*, an active verb rather than noun, to shape, mould, design, and the new philosophy of *bios* and *zőë*.

The issue of science and the body politic since Hobbes, Rawls, Bentham rights, and its relation to science, social control and democracy, has been well articulated, particularly in Shapin and Schaffer (1985) and Fuller (2010, 2011) and Gibbons on Rosseau's social contract, what it means for science. Both films used in the classroom debates open up worlds of possibilities of futures strongly linked to current sociotechnical issues in RGTs. In the school debates conducted as part of PhD research, the first question from the Wellcome Trust class activity from *The Gift* – subsequently developed for the movie *If...Cloning Could Cure Us* also – was that a character from one of the films 'was right to argue' (Jennifer in *The Gift* and Dr. McVeigh in *If ...Cloning Could Cure Us*). Students got the rhetoric of this quite quickly. While any had a normative idea about some transcendental 'right', many considered the protagonist had a *right* to *argue* based on the character's position. The philosophy of political representation suddenly becomes important, even in a school debate. In classical medical ethics arguments circulate around the concepts of beneficence, autonomy, beneficence, non-maleficence and justice. *The Gift* and *If...* both contain scenarios that reaches into these concepts. These are bread and butter issues for ethicists. But while critical to the discourses of RGTs, the term 'ethics' does not fully capture the essence of what is at stake in the practices of the everyday. The banality of the processes that have developed that may allow us to select the genetic make-up of our children, to prevent disease, set them up for the best possible start in life, or the best thing to do for society (Arendt, (1958, 2006 [1963]).

To move further into abstract territory, classrooms might explore Foucault and the 'Bodies without Organs' (BwO) (Deleuze and Guattari, 1987), life/non-life, how we see ourselves and our bodies simultaneously, all ripe for pedagogical reflection – discovery, freedom from restraints, biology.

These historical reference points in biology can all be dealt with from a rhetorically objective stance, just as the biologist does for the existing content, just as the historian can maintain a performed distance from his historical facts. This objectivity may not be going far enough, however, as we shall see in Chapter 5, when we examine the curriculum and how the clean, vaccinated objective curriculum that currently exists in most countries need not be the pure space free of ideology.

Many science education researchers from the constructivist school of thought began to see a need to teach science in a less didactic way (although *didactik* in German is more dialogic). Constructivism as applied to science education is the theory that knowledge and beliefs are accumulated as they are introduced to the learning experience of, for example, biology. It is based on the theory that we 'construct' our definitions and disciplines in a collaborative, social and cognitive way, placing empirical knowledge within a cultural and personal experience context. This field has sub-divided within the philosophy of science to such a degree – from Von Glasersfeld's (2001) near-solipsistic radical constructivism and its biological applications to social constructivism – that the definition given above barely represents the full range of views any longer. It does have some historical linkages with the type of social constructivism that Latour and friends try to escape from (Latour, 1993; 2010). The reason for this is that 'constructing' implies fakery or language games, although many of us working in this STS fields anywhere close to public engagement and education owe some debt for our philosophies to constructivism from Piaget to Wittgenstein.

Conflicts arising from discussions on objective truth have traditionally rested uneasily with the hard facts in science class, but it may be keeping alive the old debate such as social constructivism versus positivism, and introducing philosophy into science. (And in fact, the Science Wars are also a good starting point for student debate also.) We will see in Chapter 5 how ideology in science can never be removed.

Then there is Thacker's (2003) *biomedia*, the bioinformatics and codes that are a development of Deleuze and Guattari (1987). With Thacker, there is an emerging – small but influential – philosophical movement that are referred to as speculative realists, who advance, like Bennett, a world of vibrancy in non-living matter, and the 'speculative realists' Graham Harman and Quentin Meillassoux.

A further question of human body/non-body/self is emerging from the work of the Human Microbiome Project where a much larger proportion of the human body is composed of bacterial and fungal parasites than first envisaged

(Aagaard, *et al.*, 2013; see also the *Nature* 2013 special on microbiota), so much so that the older we get the more we can justifiably say we are more symbiont than self.

'Modern science' has all too easily been divided up within science education research as Nature of Science, socioscientific issues, social context of science, Conceptual change and relevance of science. There are surely connecting threads that increase the likelihood that the material of any curriculum latches on directly to Mode-2 understanding of science research, product development, history societal and philosophical concerns. (There are two recent STS stories, *Jungle Laboratories* by Gabriela Sota Laveaga (2009) and the reflexivity of education's place on biocapitalism in Clayton Pierce (2013).)

Politics is supposed to be stronger, assertive, simplified, a vision, a move that makes the difference. Science is practice with some limited inspiration, politics inspiration with some limited practice. However, let us consider how politics and science might not be such separate entities. Again let us allow Bruno Latour to enter here. He maintains, with many other STS scholars – and this is unpalatable to many – that 'the sciences' are caught up in the politics of representation, epistemology and ontology, not just ways of framing science, but ways of knowing and indeed of existing. Latour constantly makes reference to the question of 'who speaks for science?' These are not just silly word games, an acceptance that all our efforts are but representation – but we have misunderstood the connectivity of the scientific and the political, connected by a social cloud of networks and communications.

It is not 'science' at all that is messy and uncertain, but everyday practices in a particular technoscientific community (things, people, processes). This happens in a lab, separating embryos for selection, as we saw in Chapter 2. It is, conversely, the work of mediation (Latour, 1993). Purification is the work that we eventually see and hear in [the] media, as it is also for a summary report for lab managers or investors: the main points of focus, the clarity, the cleaned-up argument. Both 'separated science' and 'the impure' needs witnesses, and both types of witness are different, but they don't have to be so separate, because separation and purification interact with each other all the time.

On the subject of STS then, and the link to politics, representation and rhetoric, one important linking concept comes from Lisa Disch (2008; 2010), which has practical implications for how we link the philosophy of biology in the classroom to a future parliamentary vision. Scientists increasingly need to represent. They may well be doing the work of representative politicians already. Disch is a political theorist, influenced by 1960s radical theorist Hannah Pitkin. In Disch's reading of Pitkin's (1967) *The Concept of Representation*, she talks about how the electorate may feel that their elected representatives are not doing the job that they were elected to do. In other words they are not fully presenting the public with changes of mind and reversal of directives once in office. Disch states, following Pitkin, that those elected never followed through merely as cyphers, representing only; they never have and should not be expected to. Political representation is a dynamic process that separates itself from the electorate. They are held accountable every few years in a republic, but the republican ideal leaves them to it. So democracy need also enter elsewhere, and this is where online forums, town halls, and yes, schools come in. If we are to take Latour to the ultimate conclusion, labs also have something to say in this political forum, but these are also representatives; it is the so-called hybrids themselves, constructed from part objects, some human agency, concepts and processes that form the public, the new electorate.

How possible is this for young people arguing in the classroom? Argumentation theories tend to omit *ad hominem* arguments, as reported in our nine futures presented in Chapter 7 (for argumentation in schools, some of this ground is covered in Chapter 5). How possible is it for elected parliament? The tools are those of intellectual introspection, but the use of film and other media explores identities represented through debate – a central focus for my excursions into classrooms was the use of a popular wisdom, turn-taking, tacit practices, responding to neo-liberalist technologies and scripts, repeating lines from the script. To explore identities, we must facilitate what it is to be an adolescent (*dasein, being*), the whims, the body's oddness, the rebelliousness when no one cares (Latour, 2004b) and the identity talk can be further advanced in PGD and *real identity*, that of adolescence. We will look at these topics more closely in Chapter 4.

Anthony Giddens and Bruno Latour: witnessing and mediation

As briefly outlined in the first chapter, there are two theorists on either side of a continuum of social action used as the theoretical underpinning of this book. While sociologist Anthony Giddens is a social theorist, perhaps the most prominent in this stage of the twenty-first century, Bruno Latour is an ethnographically minded anthropologist and sociologist. Giddens has a highly philosophical distillation of his approach, and is less empirical. Latour, on the opposite end of the spectrum, has an anthropological exposition of his framework. The attributes Giddens gives to life politics is a concern for individual and collective 'self-actualisation', an evolution and extension of old-style emancipatory politics (Giddens, 1991). For Latour, facts are incomplete, do not speak for themselves, and require social, performative action. In describing performed, argumentative action in the classroom, it is argued here that a new form of politics is emerging as a result of the life decisions under discussion, what Giddens (1991) calls life politics, and that young people need to have access, and contribute, to this politics as part of their education. In a modernising agenda, for Giddens a 'sequestration of experience' (ibid. 8) has emerged, how the creation of institutions founded on biotechnology in modernity has resulted in an emptying out of uncomfortable or taboo content that deals with finitude and origins of the body. The Irish education system as a whole may be one such institution, focusing on technical skills and practices of modernity, rather than moral questions that were, and are, part of traditional day-to-day societies. Life politics then is the emerging struggle for identity by individuals and institutions of modernity to bring back into focus those moral and existential concepts once normal to Western traditional discursive practices. Giddens says this now occurs through an expression of the local/global dialectic, enhanced by media, and characterised by a turn towards life-planning and lifestyle issues of the body (ibid.; see also Butler, 1990; Foucault, 1978; Shilling 1993; Schatzki, 1996; Taylor, 1989; Turner, 1996). This turn may be evident on a global level in the words of former US president George W. Bush on stem cells, abortion, euthanasia, and so-called 'bioethics' issues, or in 'body shock' TV programming, and those topics that contribute to a society that continuously organises discourse around 'issues' (O'Mahony and Schäfer (2005).

If we were to do a kind of foursquare table analysis on Latour and Giddens there might be a clear way to see how witnessing and mediation pertains to both. Witnessing and mediation have become part of the embodiment of the person in this two-way relationship. For Giddens' pure relationships (Giddens, 1991), a particular, contradictory type of politics is in force, where there is enhanced trust *and* scepticism. For citizens in modernity interacting with expert systems, there is a 'reskilling' where

> attitudes of trust, as well as pragmatic acceptance, scepticism, rejection and withdrawal, uneasily coexist in the social space linking individual activities and expert systems. Lay attitudes towards science, technology, and other forms of exoteric expertise, in the age of modernity, tend to express the same mixed attitudes of reverence and reserve, approval and disquiet, enthusiasm and antipathy …
>
> (ibid. 7)

While Giddens sees this era of reflexive modernisation in 'late modernity', where modernity confronts itself with the sequestered getting back in as it were (Giddens, 1991; Lash, 1994), Latour famously states we are not modern at all, and have never been (1993), so the sequestered never really left the building. There are inherent conflicts between Giddens and Latour, but both strands of thinking and respective styles of thought gives us a platform onto which to launch life politics of future lives. Whatever the differences, I argue that RGTs need to be seen in the context of this risk-aware, post-normal science in a networked, sociotechnical world, a fresher understanding of the world, which we need to address as a matter of urgency. We have too easily divided progress and nature, fact and value, scientific knowledge from opinion, and we continue to pay the consequences. Both these theorists, Latour and Giddens, with a host of others, complement and challenge the politics of nature or life politics concept in society (as they challenge each other) and if this is the case for society, then the argument here is, it starts in the classroom.

There are two elephants in the room for Giddens' late modern identities: 1) can all our global villager young people afford to be fluid identities, combining public with the private, exacting change in the world, in capitalist systems where class still remains? 2) for the constructed personal identifies that take postmodern assembling, whither science literacy? Both these can be engaged with together: young people need to know what they know, and it covers more than biological content as we look at in the next chapter and in Chapter 5. Knowledge in this paradigm becomes reflexivity, more than critique but both a mediation (more passive) and a *self-confrontation* (where the institutions of science and education display reflexive modernisation and wider views of risks through the student) (see Beck, Giddens and Lash eds, 1994, with Ravetz, 2006); Mode-1 science of isolation and dissemination must be supplanted by Mode-2 acceptance. Uncertainty is everywhere, and needs to be facilitated, acted on, not ignored or countered to zero.

Bruno Latour has presented to the world, first in *We Have Never Been Modern* (Latour, 1993), this universe of non-humans with agency, a *Parliament of Things*, science fiction if taken literally – which of course in our discussions about discussions in classrooms, we cannot, yet fiction is an important device. Facts do not speak for themselves, but need to be mediated. They need the act of *speaking for*, to represent, opine. For Latour, issues that organise both knowledge and political order have at their centre *matters of concern*, rather than *matters of fact*. Again, it is 'no controversy, no issue' and the compositions created almost artistically by different potentials between actors are made up of *dings*, etymologically based on 'things', but being only partial, requiring other constructions and additional lyrics and melodies to the composition. In other words, RGTs can use the controversy and multiple perspectives in its development as an emerging technology and a way of (governing) life.

According to Whiteside (2011: 8): 'Humans always forge their identities, Latour insists, by creating relations with things'.

However:

> Latour's parliament of things creates no special standing for future generations. He calls for wider forms of consultation, but 'wider' only means: more of today's people, with whatever values and concerns they happen to bring with them. No one has a special duty to future people.
>
> (Whiteside, 2011: 8)

But Latour's Parliament of Things cannot be taken literally – how could it? How can controversy that disrupts human identity be placed into policy as 'common good' (a question for education)? Can it be a single 'common good', the 'good common world' of Latour's Parliament of Things (Latour, 2004a)? If it is to be real politics, then argumentation and persuasion must be the key methodological components of 'compositions of eloquence' (Latour, 2011).

AH Whitehead is an influence on Latour's reading of the 'bifurcation of nature'. This is a project we can use for our purposes of explaining the possible phenomena involved in future politics as well as Lisa Disch's (2008, 2010) link with political philosophy. Latour, in *We Have Never Been Modern*, does not think we should consider a world of Discourse as distinct from a world of Nature (or from Being from Society for that matter).

Then from Giddens' perspective, emancipatory politics supports a woman's right to choose; however life politics for embryo technologies, in common with future resilient society considerations, asks us to think about the scientific and moral questions of the embryo, the ethics of future procreation, the pathways and networks that bring us the point of acute decision-making. So young people's reflexivity, knowing what they know, is the work of witnessing. The expert knowledges brought through media can be used as a resource both for understanding and explanation of Western societies, as they become more socialised, mediatised. As day-to-day decision-makers – and potential future policymakers – young people are directly involved in this politics. It is a group and individual lifestyle politics (ibid.). It is a politics of identity in the sense that it concerns relationships between self, environment, human-enhancing (or -defining) technology, and how we live our lives today in late modernity. How do young people make sense of life-planning and future diagnostics? Would they veer towards perfection, like Kant's timber 'less crooked'?

As I develop this theoretically, life politics becomes a further development of Giddens, with the 'politics of things' ideas of Latour specifically, and politics of life from STS understandings (Franklin, 2000; Rose, 2007). Life politics must be understood in the new context, demonstrated by STS, of Mode-2/post-normal science (Nowotny, Ravetz, etc.), extended out to education. This type of general science pedagogy links these new understandings of the processes of 'the sciences' with education and this connecting of the STSs has two important modes and methodologies:

1. *Witnessing* This is at the core of the question of science and opinion. Here I look at new formations of opinion; once seeing was believing but now we are reliant on expert systems, as Giddens calls them, to 'make belief' (De Certeau, 1984) for us. Increasingly complex, but paradoxically fewer may be using them, these are shortcuts, framing mechanisms. The witnessing concept, with its 'Christian science' implications, goes all the way back to Thomas Hobbes and Robert Boyle (Schaffer and Shapin, 1985), the Waterford-born scientist and founding member of the Royal Society, as described eloquently in *Leviathan and the Air Pump*, the excellent and seminal science studies book from Schaffer and Shapin (ibid.). Boyle, often referred to as the 'father' of empirical science, was not content with the rational naturalism or natural philosophy of the day, a legacy of antiquity. He set up experiments with, what were at times, creaking apparatus. The air pump developed by his assistant Hooke was to become his most iconic. Boyle wanted more – he wanted a crowd to witness these spectacles (often involving suffocating birds, adding to the eighteenth-century entertainment). It was, above all else, a rhetorical performance. But it was not enough for Boyle to have an audience of peers – he wanted them to write, mediate and eventually replicate what he had just done. Witnessing moves from: 1) *direct witnessing* (first order) of those present in a 'lab', which is acknowledged as a social space, but the witnesses here are an audience of the public in wonder; to 2) *replication* (second order), where reliable witnesses of expertise wrote accounts and signed documents to the effect that they had witnessed. The replication meant they had not

only to write and disseminate, but re-do the experiments for others to see. This, many argue, is the beginnings of science. It is an automated process that brings its own mediation. Then 3) the *virtual* (third order), which has been covered by Schaffer and Shapin (ibid.) mostly, but also Donna Haraway and Bruno Latour, is the witnessing that changes minds and attitudes, as 'seen' by those, as accurately as possible, for those who are not present. While science journalists provide this role, increasingly other media do too, and much media, it might be argued, are of the second order kind (see my account of witnessing the embryoscope in Chapter 4). This classic dispute between the empiricist Boyle versus the rationalist theorist Hobbes was both the foundation and the split between science and the body politic. Hobbes needed to reflect, draw on history, theory, ideas. Boyle had to argue, to convince, get the crowd to see then convince them to re-recreate, and finally to spread opinion – not just epistemology but ontologies were created. Hobbes wanted reflection, nature seen in abstract patterns. Boyle wanted us to stare at a leaky old air pump. Both saw science as part of the political social order, but in different ways. For Latour, it is not rhetoric, nor math, nor reasoning that sells this new empirical science of the action of the Air Pump (Latour, 1993: 17–18) – as Shapin and Schaffer say, it must be the evidence of *credible* witnesses (Shapin and Schaffer, 1985). In RGTs, this would mean embryologists, geneticists, ethicists, clinicians, genetic counsellors. Should this exclude the wisdom of the crowds in internet forums, and the court of popular opinion? What the Air Pump says to other controversial areas of science is: 'You Doubting Thomases – what do you see or hear or feel is happening that is for the common good?' The senses and intuition are part of the scientific impulse, sparks of innovation. We witness an event, an occasion (Whitehead, 1920, 1927), we ascribe value, there may be intrinsic value, and this value may extend as a subject now to an object in future (ibid.)

For genetic screening lessons for young people, witnessing takes place as a film audience, as the experience in a performed debate where there is movement around the room, as a bringing forth of opinions usually not revealed (the 'gift'); many students were surprised at their classmates' viewpoints. For the embryologist and the woman patient, it is witnessing onscreen invasive procedures and the embryo to be kept, and those discarded. This is life politics. It is witnessing as learning. For Boyle, 'In natural philosophy, as in criminal law, the reliability of testimony depended upon its multiplicity' (Shapin and Schaffer, 1985: 56). Actual experiments need to be witnessed rather than thought experiments. The witness needs to give voluntary assent to the matters of fact in the experiment – not just a matter of collective viewing.

It is also important to bear witness to mistakes and failures also (ibid. 64). Witnesses needed to be reliable – the social status of the witness was important. Hobbes felt this was subversive, open to bias, what Fleck later called 'thought communities' (Fleck, 1979[1935]). Haraway's (1997) 'modest witness' is the Modern experiment-watcher, not really in 'public' as such; epistemologically public workers, but very much in a select group, behind closed doors. 'Since Boyle's time, only those who could disappear "modestly" could really witness with authority rather than gawk curiously', according to Shapin and Schaffer (1985: 26). But of course the men who worked the pump were Boyle's men. And what really was the status of the public witnesses? The modesty that Haraway (op. cit.) refers to is the objective attitude that moves out to the 'third technology', that of writing and now, peer-reviewed processes. Haraway takes Shapin and Schaffer to task for excluding the gendered aspect of knowledge production. Women often protested at the cruelty, yet those witness accounts were never taken.

There is a difference between watching and witnessing. Is it the difference between 'practice' (within discourse) and 'witness' (to step out of one discourse, or into another)? But there are also practices of witnessing, perhaps an evolution version since Boyle's time. Witnessing is seeing, attesting, standing publicly accountable for, and physically vulnerable to, one's visions and representations. Witnessing is a collective, limited practice that depends on the constructed and the never-complete credibility of those who do it, all of whom are mortal, fallible, and fraught with the consequences of unconscious and disowned desires and fears (op. cit. 267). Perhaps the witness should not be so modest? She needs to step out of discourse, and into public spaces regularly.

In life politics, when witnessing as part of bioscience, we must then differentiate between *seen life* and *unseen life*. Seen life can be imagined, smelled, felt, but most of all *visibly among peers* and exclaimed. Now this gets more complicated with witnessing since Boyle, as we move from first order, to second order to third order witnessing. DNA processes of transcription, translocation are unseen. Ivy growing around a house is seen.

2. *Mediation* The second understanding of broader science as taught through life politics running through the book is mediation. This, put simply, is the debate, the simplified discourse of a technical subject, now more likely by any kind of media, but the technologies of the performed body, and the voice, also mediate. It is better thought of as the biological medium on the Petri dish, the agar that supplies the essential life-giving nutrients. It cannot be mere information transmission – something new is always created. These are the acts of popularisation, streamlined communication framing on 'new and old' media, sociable media (Myers, 2004) planned, packaged and edited in which consumers are now also the producers; crucially there are creative elements involved for the teachers, journalists, policymakers, politicians. These are the media tricks of the trade, summarising, making shortcuts in everyday practice. These are, and should be, mediators, not conduits or intermediaries. Sources and repertoires from social,

Table 3.1 Seen and unseen life

Seen life: Human action, microscopic organisms and embryos under the microscope flora, fauna
Unseen life: Microscopic organisms described, expert systems, genetic and intercellular processes, living organs, the 'unborn'

mass and news media, games, Hollywood, conversational resources (Gamson, 1992), choosing future identities through a discursive interplay and debate, where real scenarios play out, the intended fictive, the intended factual. So RGTs takes on science and social fact, and the fictive in explanation, exploration, reasoning, debate, rebuttal, conclusion. Argumentation is fundamental to the creation of mapped worldviews and can be a natural outcome of science controversies. Although they are their own practices of the everyday, they change other practices to something smaller, sharper, clearer, newer.

Do students always have need to have 'seen', in the sense meant here, not literally, to have *witnessed* before adequately *mediating*? The conclusion that Lisa Disch reaches on deliberative public engagement activities such as citizen juries and exhaustive polling draws on what Latour had to say about public participation – there is no unmediated, pure opinion that can move from town hall to political representative (Disch, 2008 citing Latour, 2004a). But I cannot agree with Disch when she says these democratic initiatives are 'getting it wrong' as public opinion is harvested through the genres and tools of political machinery, even though this is perhaps Latour's conclusion too. To mediate, to mobilise as 'formidable actors' (ibid. 81), takes some kind of expertise that few might have. There is also of course the question of impact. However, the compositions of engagement, with their rhetoric as deliberative tools, are the only ways we have of providing recalcitrance for the 'others' at the gates of parliament. New media techniques will simultaneously make for faster inner channels to the Houses and make them more rhetorical, frivolous even. The senators in this parliament are Deirdre from the ND school speaking about her sister's PGD and Angelina Jolie at a press conference speaking about the BRCA-1 gene. The discursive genres and locations are different, but a community of concern is built up in the 'senate'.

This biopinion then becomes formation or the beginning of formation, preformations as a standing reserve (Campbell, 2011, on Heidegger) before stepping into discourse through new modes of witnessing. The practical classroom activities called on throughout the book are the tools for doing this. While Chapter 7 outlines the internalised 'product' of this witnessing and mediation, the genetic futures that are defined by the young have been validated by the students themselves and scholars. I set out, at the early days of my PhD, to do something about young people's opinions of Modern Biotechnology. It cannot be said with conviction that Chapter 7 shows young people's 'opinions'. The ideas are fleeting, changing, wrapped into the folds of the performative. And it does tell us much about the care needed with opinion surveys.

To explore this, transferrable skills/literacies can be brought into the classroom, by argumentation and consideration of biopolitical dimensions. So teachers can be encouraged to use discursive tools, particularly positioning and argumentation, in biology class. For education, these are the essential extra-curricular complexities that cause students to make systemic links, think beyond the contextual biological content, what I will repeatedly call entering into discourse. Mediation is 'responding' or 'demonstrating', the affective, the philosophical, the broadly social, even the aesthetic or spiritual, if their hearts so desire, as well as the fictive and imaginative. Life politics is *bíos* rather than *thanatos*.

There are, then, different kinds of witnessing. Seen life is easily mediated, unseen life is not. To use Esposito's immunitary metaphor further, antibodies become 'hopeful monsters' mopping up the alien life. One of Adam's (1990) human dimensions of time is that of a future colony, frozen and analysed, a time with countless billions of 'yet to be born' rather than the unborn. It is the unseen life that may not be easily witnessed.

So based on this workable assumption, let us proceed. Action is Politics. Answers are Politics. Clarity is Politics. Visibility is Politics. Blame is Politics. Bottom-line is Politics. Short-termism is Politics. All capital 'P', and performed, simplified, translated, transcended. Can their opposites be politics, small 'p', sequestered and hidden, messy, muddled and mundane, but ultimately forward-thinking and inclusive?

Note

1 Also the subject of popular fictional book by Judy Picoult (2005), My Sister's Keeper was subsequently developed into a movie, and its popularity was demonstrated by its references by young participants in the film viewing workshops during the mid-2000s.

Chapter 4

Future scripts and present bodies

Annie Kay was a sixteen year old not obviously different to many others her age. She was an aspiring footballer, athletic. Then her body began to lose control. The seizures gained in frequency – 'gaps' in her control as she called them. As she stumbles through the streets, passers-by think she is drunk. Her mother Barbara is concerned and takes her to the doctor for tests. She is diagnosed with Friedreich's ataxia (FRDA), a rare disease that causes progressive deterioration of the nervous system. The family, the test 'tells' us, has genetic predisposition towards the disease. With this diagnosis, we see familial decision-making. We never hear about a father. Annie's mother, Barbara, begins a process of denial and refuses to allow her son Ryan to get tested to see if he is a carrier or a sufferer. Eventually she does allow it. When the doctor refuses to test him for this important genetic information as he is under sixteen, Barbara becomes political herself, beginning an advocacy campaign with others for freedom of genetic information for young people. She fights the court case and wins. And so also begins Ryan's journey to being a geneticist, determined to fight Nature in his transition to the worlds of Science. Ryan studies genetics. There are decisions to be made in a crisis where susceptibility is identified in future progeny. Annie herself must adapt to a new life of difference, coping with a degenerative illness with other sufferers.

These are plotlines from the film *The Gift* (1999). The movie begins in the year 2028 when Mark, Ryan's son, discovers he was screened himself to be free from Friedreich's ataxia, but also, controversially, selected as a male with genetic markers for sporting abilities. This happened without his mother Jennifer's consent (the movie-scientist father Ryan has the ability to read computer screens full of genetic data in classic sci-fi style). Using flashback over three generations, many issues are raised: eugenics, failing health, individual identity children's rights to independence and genetic testing as well as traditional dramatic themes like jealousy, betrayal, and sibling rivalry. There is a crisis of identity for a disease sufferer and from children with labels such as 'designer baby'. Mark, the genetically selected son of older Ryan and Jennifer, is unhappy about the decision to use PGD. But why should he be annoyed? Does he not want to exist? At this concept, if discussions were on today's internet boards rather than mid-2000s classroom, the students might respond with: 'WTF?' And what of the other protagonists in the network and their other connecting networks? Barbara, the mother who wants her son to be tested for Friedreich's ataxia; Annie herself and her support groups, advocates for better care in society for FRDA sufferers; the older Mark's decision, the embryologists and technicians, presented as objective and machine-like in the lab, the colder genetic counsellors in the clinic.

This chapter outlines the real-life decisions of patients and families interweaved with school classroom audience, fiction and aesthetic worlds[1], invoking a cultural studies flavour on the overall STS study of RGTs in education. There are strong practice-based connections between film representations, active audience debate and real-world decision-making, some of which is taken from Irish fertility clinics. The biopolitical dimensions we explore in the previous two chapters are viewed here 'on the ground' as it were – how schools and students react, based on the PhD study carried out between 2002–2007.

We briefly zoom into the lives and practices of some of our participants in six schools, students and teachers, while interweaving the narrative of the two films shown in class with those of the participants. They are the plotlines and scripts of life-planning – the audience and the presentation on media – the factual and the fictive enmeshed. These are the lived experience and the cultures behind the debates in classrooms, ideas that may one day lead to real life-planning decisions.

The schools that participated in this study offer empirical evidence as supporting case studies for resisting or accepting a position on strategic technologies, caught up in stories from science fiction, the curriculum, or the newsroom and filtered through or sequestered in the classroom. This chapter begins the argument – given the cultural/strategic restraints of Chapter 2 – how can our own 'characters', the students and teachers, deal with hard borders at the edges of the classroom? The solution is to relax the border controls, allow a pedagogy of external narratives to filter through, on cloning, biopolitics, identity and indeed environment, the broadest use of the term. There are many narratives, or cultural future scripts, to borrow a phrase from a future nanotech innovation STS team in Amsterdam (de Laat, 2000; den Boer *et al.*, 2009), which crisscross our lives in many ways. For young people in a twenty-first-century biology classroom there are competing interests. I want to switch from screen to classroom and back again with more detail to our story's characters, factual and fictional, to show the performed connections when talking about future children and future genetic ties. There is a little ethnographic detail of students, teachers and 'outsiders scientists' dealing with the boundaries of the science subject, film, work, fun, identities, futures, Science and Nature. The classroom arguments are reported in more detail in the next chapter.

There are two spatial conceptual terms of concern: i) *sites* are contested conceptual spaces represented as real objects in the world: bodies, schools, teachers, students; ii) *locales*, from Giddens (1994), are once-traditional physical geographic spaces – but like Bourdieu's fields, the wheels of an engineered modernity have lifted off and disembedded from these locales, so their visible form stretches over geographic space (and informed by the alternative, but similar, STS-geography readings of locale of Thrift (2005; 2008)). A locale can be a conceptual, but localised, space – a space of common practices, but where science fact and fiction enter from outside, where information from the Beacon reproductive clinic and *Gattaca* (1997) become part of the structural representations of RGTs interacting with young people's own representations.

While STS and science education offer insight into the practices and structures, and indeed lifeworlds, the approach of cultural studies offers new qualitative insights that show us more about how young people will respond to ideas about science, fiction and authority, none more so than the focus on *identity* and *identifications* of Stuart Hall. I have, therefore, been keen to keep a cultural studies flavour to this STS project. Identifications here can be both with onscreen fictional characters and people known to participants, with characters, but also identification as Hall describes it, a projection of more fluid identities, a social process, an interactive, performative set of identities formed with other peoples and ideas, an imaginary of plenitude (Hall, 1996). It can also mean the concepts and identities that we as educators and researchers create for them, and they as students may resist in the context of biology education as they construct moments of their own self-identity. Identity choice is seen here as a process of discursive action, identity as voice, following Goffman's (1990) [1959] performative practices and framing strategies (ibid. 1975) and Giddens' (1991) authenticity of self that creates a new flexibility – and reflexivity – for moral self-identity in modernity. It is 'the creation of constancy over time, that very bringing of the past into conjunction with an anticipated future' (Giddens, 1994: 80). The cultural, and indeed biological, identities of young people become important sites of action in a risk society where they are asked to negotiate through education, through life-planning and careers, through trust in expert systems, through life itself – to learn about their bodies, environment (as Nature) and futures. Lash's astute analytical difference between Giddens and Beck might suggest ways with which the young might deal with this: while Giddens sees the structural worlds of expert systems as sources of personal reflexivity – the inner working of an iPod, the processes of clinical care, the car wash, the setting of a curriculum – Beck expects that we must escape them to be truly reflexive (Lash, 1994: 116). So while Beck's risk society is a powerful explanation for our constant awareness and ordering of risk discourses and information, Giddens offers the best way forward here – all the genetic information on disease and characteristics, constructed from conception, all the biological knowledge, entwined in popular culture understandings of the self, the mediated environment (literacy here is a different thing than that which is read from a biology book alone) are part of an authority to be learned, and with authority comes tradition, running up against the modernisation project of identity in a kaleidoscope of ways (Giddens, 1991).

The idea of future cultural scripts are an important way of mapping out what is important to the contributors in a social science project. Chapter 7 examines these in more detail; meanwhile, to ensure we see the interacting and co-producing of structure and agency, we will here examine the stories of the past and present and how they lead into future speculation about RGTs, life, livelihoods, and future bioscience/life governance.

Positioning and performativity: bodies in trouble

The schools that participated in the early study (2005–2007) offer empirical evidence of the resisting or accepting of a position on strategic technologies, caught up in stories from science fiction, the curriculum, or the newsroom and filtered through or *sequestered*, to use the Giddens (1991) term, in the classroom. And here is the main argument: how can there be such hard borders at the edge of the classroom? We look closer here at these contested borders, how embodied action of teachers, students and the odd visitor helps to pull the macro issues of genetic technologies, identity, future-planning and fiction into education.

Identity and embodiment are rarely captured for the science communication part of education, but nonetheless they are important expressions for witnessing and mediating, our two central pillars of life politics already elucidated. We can discuss here how current bodies are *presented*, but also the body (or in our film, a baby) as a *present*, the notion of gift as a *motif* from Mauss to Esposito (2008) (the 'temple of the body'). But we are not interested here in the 'gross bits' of adolescence, coming to terms with their bodies and emotional interactions, the realms of sex education (RSE), and sexual politics. This is human reproduction without sex, but also reproduction of another kind, not entirely different, the reproducing of assumed knowledge and ideologies form teacher to student, one generation to the next, from Bourdieu (Bourdieu and Wacquant, 1992; Bourdieu and Passeron, 1977). It is also the clichéd gift of life through childbirth, or external body kinship and familial ties, now also possible through IVF and PGD, a Body without Organs (Deleuze and Guattari, 1987). In an early scene, Mark receives a printout. He is sixteen, and in this sci-fi universe, he is at the age of receiving a readout of his genetic profile. This printout, a rite of passage, propositions us – Mark, his peers, the audience – making the strong claim to *be* Mark's identity. But the identity does not arrive. Mark is confused and concerned to realise that the printout is essentially blank, due to 'fixing' by his geneticist father Ryan to cover the clandestine act of positive selection. Mark's friend jokes: 'You could be a figment of my imagination'. The philosophical conundrums of genes, brains, vats, *deus ex machina* and nature are only raised in this one scene.

If we are to keep Giddens in the picture, then structuration is key, the interface of structure and action – the ideas, language and narrative of the body versus the embodied action itself. For Latour (2004b), the proposition into the Parliament is articulation, body as language; this is more pragmatism via his reading of James (1996 [1907]). However here is no doubt, in contemporary theory, since Butler (1990), Billig (1987), and Shilling (1993, 2005), the body has claimed central place. Beyond the body, there are many narratives, or cultural scripts, to borrow again the phrase from Arie Rip and his Netherlands colleagues (de Laat, den Boer *et al.*, 2009), which filter into our lives in many ways. For young people in a twenty-first- century biology classroom, there are competing interests: Facebook, YouTube and other social media, computer games, fashion, music downloads, film, TV, box-sets, being cool, staying noticed, girls/boys, maybe getting wasted. The body is the social site where these things still happen (although many of the things listed in the previous sentence appear to indicate a move online or further into ephemera, apparently disembodied). In a dangerous online/offline world, teenage bodies are made up, pierced, tattoed, dyed, tanned, shaved, haired-up, starved, fattened, inebriated, beaten, and sometimes even allowed to sleep; they may be then photographed (perhaps as a selfie) and tagged by smart phone for Facebook and geo-tracking, then trolled and LOL'd. As much as these disruptive, performative events require the body, they would hardly occur in the classroom. In the classroom, the body is forced into a seated position, behind a desk, uniformed, slumped and disinterested, or alert and attentive, in a system that Foucault would claim is a type of biopolitical imprisonment (Foucault, (1995 [1977]). It hardly seems a fair comparison, but in Foucault's typology of institutionalised discipline in traditional and contemporary societies, the school and technologies of the social space rest at one end with prisons and mental institutions at the other. Students read or write, but are rarely yet asked to use handsets or iPads; certainly not seven years previous to writing this when data was first collected, although currently changing in certain schools. McSharry calls them schooled bodies (McSharry, 2009), where she draws attention to the understudied realm of the embodied sociology of adolescence and peer structures.

These online options were not so available in the mid-2000s. If the secondary school discussions and debates were to occur at the present time, these teenage groups would have ask.fm (2013) and spunout.ie (2013), controversial websites that provide information on sex, suicide concerns and emotional issues. In 2005, Bebo was just the beginning, the migration online for youth identities.

Both content and spaces of education have been explored (see for example Tobin's *Good Guys Don't Wear Hats*, 2000) but rarely has the content of contemporary biology (and its implications) been studied in this way. Studies of bringing what we know into the secondary school room from STS and cultural studies are remarkably absent, although there is research on performed bodies, style, externalities (ibid., Giroux, 1992; 1994; 1997a; 1997b; McSharry, 2009).

Science fiction offers the intersection of science and art as popular means of engagement. But it is more than a dissemination exercise or the use of science for conceptual change. It is complete immersion and engagement for the student with the structural issues of biopolitical change. There is physical bodily movement too in the exercises I have used, when students are asked to move around the floor and position themselves in debate. There is an exchange of the embodied movement and screen. Media provides a vital area of learning and experience for all students, the proximal mediated, issues of changes to the body, the enhancement, the plastic surgery (see Reiss and Caccavale (2008) on young people and monsters in science), represented on such TV programmes as *The Swan*, *Body Shock*, as well as the more elite forms of bioart (Nelkin and Anker,

2004). There is aesthetic richness in the perma-changing, species-defying cyborg bodies of Orlan and Stellarc, the gender-morphing of Genesis Breyer P-Orridge, the bio-engineered animals of Eduardo Kac. This is a new commoditisation as we refer also to young people's 'Commodification' future frame in Chapter 7. The body is constantly objectified through history, public executions and the public autopsies of Andreas Vesalius (Sebald, 2002) to the prevalence of pornography. The objectified body, and in the twentieth century, the commodified body – organs, molecules, genes – are now objectified into the several economic processes (Rose, 2007) and coming with news of bioscientists with aims to have Frankenstein-like engineered spare parts. The obvious contemporary example today to Vesalius is artist Gunther Van Hagens. Both Habermas (2003) and Esposito (2008) ponder at the potential relationships with future bodies that can be easily changed with variations of flesh, synthetics and genes and how 'lesser humans' or lesser consideration for human form could result in an immunitary (ibid.) or commodified objectification (op.cit.). The movie *Gattaca* (1997) shows us a lower caste of people who cannot afford genetic screening or engineering. 'My destiny was mapped out before me – all my flaws, predispositions and susceptibilities – most untreatable to this day. Only minutes old, the date and cause of my death was already known' (Jermome, in *Gattaca*, (1997)).

These are all part of the ethics of a kind of moral bio-enhancement, as we pondered in Chapter 2 when we analysed the ripple effects of thought communities and the radii from bio-orators, bio-governance as social order, social scientific political control tightly bound to biological progress. Is a soma-controlled citizenry really that desirable to its (democratically elected) rulers? Whatever about biopolitical control, let us assume bio-robot subjects in liberal democracies are not the desired outcome.

So let us ask again: why study RGTs as a subject among young people? McRobbie (1994) reflects on the relevance of using young people's views. The glib, pop culture, good-life answer is: the children are our future. In more scholarly terms, in studying late modernity, young people offer a variety of social, historical, and psychological insights. According to McRobbie: 'Youth remains a site of cultural innovation' (ibid. 179). If there are new ways of looking culturally at RGTs or a modest change in old ways of knowing, the young provide some valuable insight. Henry Giroux, the critical pedagogy scholar, wonders why educators are not so concerned with these cultural matters of the classroom, 'the politics of their own location' (Giroux, 1994: e1). As raised further in the next chapter, the same question could be asked of science education researchers. The realist answer to this question of course is, these young people are close to being young adults and any may face these RGT decision-making dilemmas much sooner than anyone dares to think. The curriculum does not deal with such transformative, biological-biopolitical change adequately. Indeed the academic discipline of science education tends not to deal with subjectivities in the science classroom with exceptions in Aikenhead (1996), Cobern (1996), Costa (1995), and Phelan *et al.* (1991).

We need also to be aware of how the young represent themselves in such discourse. Framing strategies, described in Chapters 6 and 7, that is, making collective sense, or playful nonsense, of an issue within a group, are also part of the resources of the young as they negotiate their own identities while contributing to the discourse of RGTs in education at a local level. These discussions are brought into pedagogy, including schools in Ireland run by religious orders, places where it may be expected that some tension exists when expressing these views. Zooming closer to pedagogical action we see the illocutory speech acts and positioning (Harre and van Langenhove, 1999) and even what Latour sees as the speech impedimenta (Latour, 2004a), the inarticulate of not just young people, but of use all in the face of such profound human biological change. We all can learn the *mythos*, *pathos*, *logos* and *ethos* of Cicero, Goffman's (1959) back stage and front stage framing, with science itself on stage and communicating outwards (Hilgartner, 2000).

For Giddens, following the Butler trajectory of embodied identities, making a body more like oneself, or a truer self, is another type of subject-object paradox – the potential to use future technologies for more authenticity, like Heidegger's nearness, as a means of being more oneself while literally shedding an old body is part of the constant switch between flesh and Deleuze and Guattari's BwO, a greater fluidity of self-identify between what once was a true body and what is now a truer body. Cosmetic surgery, prosthetics, the familial relations of parent-progeny demonstrate that the body is increasing more a site of cross-action, a site of self-promotion. This type of outward projection is a proposition position to Latour, requiring constant socialisation. Following Stengers, he says: 'The word "proposition" conjugates three crucial elements: (a) it denotes obstinacy (position), that (b) has no definitive authority (it is a pro-position only) and (c) it may accept negotiating itself into a com-position without losing its solidity' (Latour, 2004b: 212). We might also say 'pre-position' as students respond to difficult, perhaps poorly conceived, concepts. They have yet to experience family planning *or* lab science.

Snapshots of locales and spheres: media and science fiction in informal education

The EC guidelines on science education are moving slowly towards Mode-2 thinking, demonstrating a twenty-first century full of perceived risks. However, how different is this – is pedagogy just add-on theatrics interpreting the same curriculum? Levinson (2010) proposes a mix of art, identity and citizenship for science, as Rose (2003) did – quite influential for my PhD research – by introducing the topics of sci-fi clones, dead people and *The Fly*.

I introduced locale in this chapter as a place, but it is more accurate for Giddens (1994) to call it defined and confined *space*, disembedded as it is from locality. Thrift's (2005) locale would privilege place and biology together where Bourdieu's action occurs, the *habitus* constructing a subject, the life-planning of identities of Giddens (1991). Social theory, science education research and STS all intersect here. And so much is taboo in the classroom, sequestered as Giddens says. Finitude is not allowed in – there is the topicality of teenage risk of suicide, but for all this, education hides away from death, or embryos, clones, euthanasia, multiple parent commissioning of children.

For Henry Giroux (1997a, 1997b), gender is dichotomised with ever-increasing rigour from the time children are born. The girls are brought down a pink pathway of passive acceptance of hair accessories and with boys far away. The boys are challenged more, inculcating science and engineering. In advertising aimed for boys, the audience (toddlers to pre-teens) are asked to construct, muck about. Girls of the same ages are told to present, to order for social acceptance. Drawn in by market forces, Giroux calls it Disneyfication, and we can see it clearly in the Disney Princess pull of young girls (ibid. op.cit). This suggests that designer babies are already with us, and, through the structured external strategies of the market forces become designer tweenies. Critical pedagogy has the intention of challenging young people to see these forces, to help young people to create for themselves an alternative life-plan that does not get caught up in neo-liberalist agenda, as they pay for what they see, and what they *are*.

There are perhaps inferences to be drawn here to the work of Couldry (2003) and Turner (1974); ritual as a process of making defined boundaries in media, 'framing the social', or media events and people *presented* in a media ritual but *representing* society.

As we talk of 'screening', ironically there are few screens in contemporary biology classrooms. In the early twenty-first century, the screen is our way of first, second and third order witnessing, yet we need our young people to face other objects only, including the teacher's oratory. In *The Gift*, there is an over-aestheticised laboratory, all futuristic and *Star Trek*, with DNA bubbling and twisting up along screens (the film is pre-*Matrix*); the clinicians are cold, detached, rational. This contrasts with the embryologists at Beacon CARE Fertility, a commercial clinic out in 'the real world', who demonstrate positive personal feelings about the ethics of their actions and the good life of their patients. Unlike *If…Cloning Could Cure Us,* the other movie used in classroom discussions, the scientists in *The Gift* are somehow part of the non-human equipment, the bioscience structure, making logical and rational decisions. Latour and others have well demonstrated that the technicians and embryologists are indeed part of a greater composition of objects and equipment of technoscience, but are neither cold-hearted nor objective in how they bring the findings to the world. They are human witnesses, spokespeople to laboratory action (ibid. 1993). Jennifer, Ryan's wife, is not present to witness the key turning point in the film where Ryan and the fertility clinic assistant (both male) conspire to choose a male embryo with propensity for sporting abilities while watching the sci-fi screens of DNA. In Beacon, the mother watches in situ during the taking of oocytes, and the transfer back into her uterus of selected embryos. In between, there is a second-order witnessing, a transcendent moment of witnessing for parents at the embryoscope. On a PC with HD screen, prospective parents can see in, time lapsed movies, all the embryos with cells dividing in the IVF or ICSI process.

Most young people are familiar with blockbusters where there is some form of genetic control or enhancement, such as the myriad of Marvel comic book heroes. From a social order perspective, possibly the best-known movie is *Gattaca* (1997), which deals with a two-tiered genetic society, those that can afford RGTs and those who cannot, and who become instead Agamben's *homo sacer.* Such films can be shown to a biology class as sci-fi dystopias set up as the classic jeremiad – if we do not act, or change, this is what we will become. But there are, as we shall see, much more nuanced readings.

When I first played films for young people as part of biology class, there was a strange resistance when fitting the activity into what science may be. Media studies might be seen as too much of a good time – how could this be learning? There are two main protagonists in *If...Cloning Could Cure Us* (2004) – the 'maverick scientist', Alex Douglas, who decides to harvest stem cells from a 19-day-old embryo (the law states no older than 14 days) to develop treatment for the victim of a climbing accident, and Andrew Holland, the victim himself, who is wheelchair-bound. There is a central courtoom scene, and connections between Alex's actions and the global contemporary networks of bioscientists and ethics.

As I write this in 2013, *If...* has dated the least of the two movies. While *The Gift* is obviously an 'edutainment' film designed for classrooms, and BBC Bytesize, *If...* is in dramadoc format, made for Channel 4. As I recalled in Murphy (2008), *If...* is more polished and clearly has had more funding put behind the project, with expansive exterior scenes, pumping soundtrack, clever editing. It also used in quite an interactive way, with alternative endings for the trial of Alex Douglas when it was first broadcast in December 2004. A guilty or innocent verdict was played after an audience vote. There were 750,000 viewers, which represented 3.1 per cent of the total audience share (Reid, 2012: 987).

Reid (ibid.) carried out a comprehensive study of the use of this particular movie, with twenty focus groups. Y-Touring has an earlier version of *The Gift* through the Genetic Futures website, now updated by Theatre of Debate (2013). For Reid, the dramadoc is a rich, new type of science communication (ibid.). But more than the communication of science, both these fictive examples have used cross-media *agora* to not just inform, or entertain, but stimulate involvement. Fiction, genetics policy, global bio-trafficking, courtrooms, witnesses: a future, interdisciplinary Parliament of Things (Latour, 2004) would surely draw on such mediated politics. Film and fiction can set the scientific legitimacy of Ryan and Alex, the scientists in *The Gift* and *If... Cloning Could Cure Us* respectively, in opposition to the kitch of *The Island of Dr Moreau* (1996), the adult horrors of *Splice* (2009) or subject-melding of *Upstream Color* (2013). Popular science literature with philosophical ideas about human biology and 'non-natural nature' (Bucchi, 2004, citing Haraway, 1997) fascinate as well as frighten the literary imagination. Reiss and Caccavale (2008) demonstrate how to bring such horror safely into the classroom without fear from the teacher. Zombies and terror plagues have provided rich pedagogical impetus for bioethics and other globalised, new order values while always retaining the aesthetic and rhetoric of pedagogy itself, a crucial approach (Brummett, 2013). Bioscentists may be postponing or eradicating future death, but it cannot be eradicated or sequestered from the pedagogy of life politics. Fiction helps to set it up as distant morality through mediation (Boltanski, 1999), bringing the messiness and sequestered back in, while keeping the environment of the child safe.

The choice of *The Gift* characters are life choices, and they are in networks: Annie's decision to join a patient support group to deal with her illness; Jennifer's decision for natural – random – embryos to be selected free of FRDA genes; young Ryan's decision to be tested, to become a geneticist years later, and the key decision in the film he later makes – to select a male embryo with genetic markers for sporting abilities. In *If...* the big decision is Alex's when she works on a 19-day-old embryo rather than a 12-day-old one which is within the rules. But then the distant morality of film and media opens up other whole networked worlds, other compositions. In *If...* it transpires through courtroom cross-examination that a Chechen woman was the donor of the original eggs used for the cloning of stem cells. She was the victim of black market dealings; the 'other' world of poverty, this woman was forced to sell not just her eggs just to make a modest amount of money to survive, but in the brutal procedure used, her uterus was also removed. This ties in with legal accounts of black market organ smuggling and other bio-unethical practices, in what Widdows (2009) concludes is subordination of dignity and identity on grounds of gender and indeed is another type of subordination that includes all reproductive and labour discrimination of women globally. A woman, in a country with no connection to the central RGT, is forced through poverty to sell her eggs to provide the stem cell research procedure that has a limited probability to make an injured Irishman walk again (and there is a risk he may die), a man paralysed following a fall while engaged in a recreational pursuit. The uterus is taken from the woman as well as the egg, and with it, an integral part of her female identity, and so ironically, she cannot have children herself. Hadjat Tataev from Chechnya is, in the compositions of global biopolitics, *zőë* (Agamben, 1998; Esposito, 2008). Again, this goes beyond bioethics. This lays out a terrain of global compositions (assemblages as Ong might call them), the battlegrounds on which life politics plays out, that sits alongside other contemporary (but not 'Modern') topics such as corporate social responsibility, ethical trading, environmental and emancipatory ethics and even healthy lifestyles[2].

An interesting phenomenon that occurred in classroom sessions when showing the film to students related to the internal structure of the film. Once we enter the discourse of decision-making, what was represented onscreen sometimes becomes left behind, with loose ends tied up, and off-screen actions and events filled in. Participants begin to ponder on what characters might have done and motivations that exist beyond the depth of this limited forty-minute drama, much like fan-fiction.

As covered in more detail in Murphy (2008) when relating these sessions, it seemed that the film script leapt from onscreen to the everyday lesson scripts of communal speech act and back again. Tobin (2000) calls this type of discourse *citationality*, sometimes involving re-speaking of lines belonging to others. By replaying a line that is heard before, citationality is a close crossover between copying a line from some media memory or what we will see occurs in Chapter 6, media theorist William Gamson's connections of experiential material, mediated sources and popular wisdom, and the confusions in recalling where data was sourced (Gamson, 1992). It may be unclear how such a line became replayed again and again as part of the cultural roundabout of words in the constant deference of meaning (Tobin, 2000 citing Derrida, 1976), whether it was media derived or overheard in the past during a conversation.

This invokes the dramaturgy of Goffman, bodies are 'used' in discourse, as well as the well-wrought front stage/back stage analogies of interpersonal communication. Young people – everyone – in social events are both human and social bodies (Schatzki and Natter, 1996). Judith Butler (1996), invoking Bourdieu's 'social magic' speaks of rules of *habitus*, the social structure inscribed in self, reacting to its structural presence through practical mimesis. The *habitus* – embodied history, internalised as a second nature and so forgotten as history – is 'the active presence of the whole past of which it is the product' (ibid. 45, quoting Bourdieu, 1990: 56).

Even the serious play-acting of young in teenage love-struck or awkward or rebellious mode cannot avoid the political. They are body-complexes acting and reacting, entering into moral discourse across a lattice of practices to deal with wider social issues of identity (Davies and Harre, 1997). This is a broader view of pedagogy, the boundary interactions between school, media, institutional technoscience, and the young citizen that operate around the curriculum. The geography of school, classrooms, timetables, subject disciplines, school uniforms, and so on all contribute to structural constraint in which the student agency must negotiate when accessing bioscience information permeating other spheres of life.

Mimesis is the act of repetition, of copying, sometimes peers, sometimes celebrity or screen – maybe in moral mode, parents. Gabriel Tarde is another 'losing bio-orator' of sorts; the history of sociology has decreed that, in the debates with Durkheim on social order and action, Tarde loses out. Tarde's sociology of biology too, is a conservative one, with constant repetition, seeking of opposition and adaptation. Latour has brought Tarde's thought back into the limelight. Innovation is the creative spark, pre-positioning, the stepping into discourse. In Nietzche as with Tarde, there is a higher structural force happening with mimetic action (ibid.; Lash, 1994). In the contemporary, fuzzy sense of the term, can biology allow innovation?

Chapter 5 recounts many occasions where students expected rote learning from science and learning generally, and were confused by this sudden departure into film and debate. For SKt, there is a constant requirement to follow the lesson plan. Addressing students' ethical questions are problematic:

> As a teacher you have to be conscious of [the] people in front of you. You may have somebody who has an inherited [disease]. Like for example the other day there we had the person with a lung [problem]... But you may be conscious that people. ... you're talking [to have] insulin deficiency, you may be conscious of that. So you have to be conscious of the group that's in front of you now. People are well aware that if you make any sort of a statement and it was in any way inaccurate... You know. People now can access information from the internet, from at home, so you have to be very careful that you don't give out [inaccuracies]. You almost stick to a prescribed script. And that is something coming more and more into play. So y'know the cohort of people in front of you, different backgrounds, people have different values now and you don't ... and you don't give your values to the group. So you have to be open to that.
>
> (SKt, interview)

In this passage, SKt has, through a range of concepts, identified the core issue of power and performance in classrooms – the fear of competition for facts. He feels he needs to be a performer to deliver these facts. He needs to 'stick to the script', a script that is objective, accurate, scientifically sound, and neutralised of personal values to compete with external 'experts' delivered through increasing media imagery. The reason dramaturgy is important here is that the students' performance tends to overpower the content, although a balance can be struck, when one gathers oneself, to enter into discourse. This applies to a class presentation or debate as much as a media interview by a PR-enhanced professional, who has four minutes to put her three points out there.

Snapshots of sites: the schools and participants in *The Gift* sessions

Before we explore some ethnographic detail of students, teachers and 'outsider' scientists dealing with the boundaries of the science subject, film, work, fun, identities, futures, Science, and Nature, let us consider the idea of a *site*. While locale, as we have seen has been disembedded somewhat from geography, so too site, using as I do the Ted Schatzki (2002) understanding of it, is increasingly social, but can be more localised than Giddens' locale (see also science fiction's role in mapping out future spaces in Kitchin and Kneale (1991). Esposito (2008) sees something newly malleable, newly contested in the human/nature hybridity since Foucault. The biological human/nature hybrid is, to him, also a site, pushing resisting being a cause, a force, an action, a reaction. In some cases, I detected, with sadness, hints of the 'lads' in Paul Willis's (1981) classic *Learning to Labour*, young boys caught by fate, a determinist education system that prepares them as they gloriously rebel, for the hard, unskilled labour of their fathers.

The sites were six schools participating in the core analytical phase of this study. They were located in the province of Leinster in Ireland, namely North Dublin,

South Dublin, North Dublin city centre, South Kildare, South Wicklow, and Louth (Table 4.1). They were selected to represent a wide demographic: urban (ND, NDC); rural (SK, SW); isolated from the local community (NDC, SD); high number of international students (SD); affluent background (SD); disadvantaged (ND); all male (NDC); all female (LH); expected Catholic/religious ethos (NDC, LH); and a wide geographic spread within the province (LH to SW). Other schools were involved in a pilot project from 2003–2005, to demonstrate how a pedagogy of life politics might exist for those outsiders beyond the formal

Table 4.1. The participants during the recorded sessions in this study

ND	North Dublin; disadvantaged area; fifth year
ND1	'Deirdre'
ND2	'Donal'
ND3	'Denise'
NDt	teacher [female, middle-aged]
SD	South Dublin; boarding school; affluent; mostly non-national students; fifth year
SD1	'Della' [from Germany]
SD2	'Daniel' [from Bahrain]
SD3	'Douglas' [from Kuala Lumpur]
SDt	teacher [male, thirty-something]
SK	South Kildare; village school; farming community; Transition Year
SK1	'Kevin'
SK2	'Katherine'
SK3	'Kate'
SK4	'Karen'
SK5	'Kei' [Japanese exchange student]
SKt	Teacher [male, nearing retirement]
SW	South Wicklow; village school; farming community; author's alma mater; fifth year
SW1	'Wayne'
SW2	'Will'
SW3	'Wanda'
SWt	Teacher [female, early middle-aged]
NDC	North Dublin City Centre; all male; affluent; Catholic ethos; fifth year
NDC1	'Derek'
NDC2	'Des'
NDC3	'Damien'
NDCt	Teacher [female, twenty-something, first job; work placement at NICB]
LH	Louth; all female; Catholic ethos; Transition Years
LH1	'Lorraine'
LH2	'Laura'
LH3	'Lisa'
LHt	Teacher [female, middle-aged]
Bioscientists	from NICB
Presenter A	at ND, NDC, and SK.
Presenter B	at SD, LH, and SW

Note: Where these names and codes appear in the text, please refer back to this section. Teachers and bioscientist presenters are referred to by code only; Presenter A and Presenter B were the bioscientists and LHt, SWt, NDCt, NDt, SKt and SDt the teachers.

education system. The participant ages from the research schools ranged from fifteen to seventeen years, from Transition Year and Leaving Certificate classes.

The task of linking the cast of characters who participated in the PhD data collection of the study with locations might seem daunting to the reader. To facilitate the identification process – while still retaining anonymity – schools were identified in the original PhD study by two or three letter codes that reflect their location, while students were referred to by false names beginning with the first letter of their home location. 'ND' then is North Dublin, 'SW' was South Wicklow, 'SK' was South Kildare, 'NDC' was North Dublin city centre, 'LH' was Louth, and 'SD' was South Dublin. Deirdre and Della were from *D*ublin (north and south respectively), Kevin from County *K*ildare, Lorraine from County *L*outh, and so on. Each student also had an abbreviated code name such as ND1 or SK3 for brevity but the pseudonyms were more often used when referring to them in the text to personalise them.

The codes further served to connect participants to a place, e.g. SK = *South* Kildare, ND = *North* Dublin. This system also greatly aided my own memory during analysis, although there is a complexity about using a real name, false name, and code.

We will now look more closely at particular events and daily practices from the cultural surroundings of the participants from the secondary schools, such as the history of a school or its acknowledged ethos, often told in the words of the participants themselves.

ND, North Dublin

Part of my early analysis of the cultural placement of the young within controversial RGT issues had been to identify Costa's (1995) outsiders to science. ND was one such school, considered by the then Department of Education and Science as a socially disadvantaged school. Local authorities and social inclusion network-supported initiatives had encouraged young people to regain a sense of pride in place, and to represent this artistically. As I took most notes after Christmas 2005, these same artistic imaginations were on display in the SK foyer. Large posters decorated the main atrium, representing fun aspects of New Year's resolutions. In among scientific-themed posters leading from the atrium to the biology lab, NDt had placed an *Irish Times* article on the discovery of *Homo floresiensis*.

Deirdre and Donal, two student interviewees from ND, described the negative self-identity of the community. Deirdre is fed up with what she calls the 'stereotypical' views of people outside the ND area – particularly in the country – who, when they think of her home, see only crime.

I: They're kind of afraid of you. They kind of like go 'Oh you're from.....*that* area. We ... won't mess with you' like.....Even in jobs or whatever, they're always keeping an eye on you and stuff. So I think it's kind of hard to get a job like anywhere else if you're from this type of a school.

PM: What do you think yourself?

I: I think it's brilliant. I think the school does a lot for the surrounding area. Like it gives you lot of hope for our future like. Better education and stuff.

(Deirdre, ND1, interview)

Donal also thinks that people from outside the ND area have a narrow view of it. 'There's bad apples everywhere,' he told me. 'There's just a few … that give the place a bad name,' (Donal, ND2, interview). In this case Deirdre and Donal present the surrounding environment of ND as an outsider *area*. This in no way means they would define themselves as outsiders to science. Denise, on the other hand, is an outsider within an outsider area, or a marginalised area as perceived by the inhabitants. NDt tells me that Denise is one of the weaker students and thus, in her view, might not be suitable to participate in the study.

Denise, in common with Derek and Des from NDC, describes 'going out' as a hobby or interest outside of school. Denise however makes interesting references to the 'small details' that come up in biology class, such as drink and drugs, and these shine a chink of light on her peer culture. A particular sort of relevance emerged with emotive references to home life that connected with the film or bioscientist sessions. Deirdre said: 'Mam had breast cancer…Dad coughs a lot. A heavy smoker' (Deirdre had been interested in DCU Presenter A's description of smoke entering lungs and causing tumours).

The young people from this area sometimes have to juggle school work with paid work outside of school hours. The teacher, NDt, remarked to me that they might often resent having to come to school when they are not getting paid for it. This was the norm during the Celtic Tiger years. Deirdre worked in the half-term break for a pharmaceutical company – that also employs her father – 'scanning stuff and tak[ing] orders for hospitals and stuff like that'.

All three interviewees gave answers which were slightly more flexible than their rigid positioning when they performed in debates. In interviews, they were inclined to stay close to the centre of agreement – that is, undecided – on whether to use RGTs or not. Closer examination of the RGT issues on a one-to-one basis revealed more common ground between participants than in the forum of debate

Deirdre was a 16-year-old mother who described herself as 'stupidly funny'. She enjoys 'hanging out with…. friends' and 'drawing'. She is a quiet, reserved person. It emerged during discussions that she has a niece with a cleft lip, with whom she identifies ('because of the teasing'). She also identified with her sister, the child's mother, who might use PGD for future pregnancies. She was open and honest about the child's condition and her struggles in school. She is mature beyond her years and in her responses expressed a deep sense of protection towards anyone considered physically 'different'. Deirdre expressed a view in class that parents have to make tough decisions, including embryo selection if this becomes a form of 'protection' against disease. She uses deadpan humour in interviews, and although her answers are generally short, they are full of information. She referred a lot to the film and told me she was interested in pursuing film studies.

Eight years later, I was reminded of Deirdre and her sister as I was shown around Beacon CARE Fertility patient waiting rooms, located on the more affluent south side of Dublin. Deirdre, it appears, was mistaken, or there was some miscommunication. The single gene TP63, associated with Ankyloblepharon-ectodermal defects such as cleft lip or palate, was identified and published in the early 2000s (Sutton *et al.*, 2010). There may well have been discussion in Deirdre's family about reproductive options in 2005. There would have been more options available in the UK. But only one case went through HFEA since this publication, and this is seen by many as a step into social, eugenic screening, as there is surgery available for cleft lip or plate. For whatever reason Deirdre's sister needed advice on PGD, were she requiring the service and able to afford Beacon for its range of options, she would have started with intensive consultation.

Front stage, staff at Beacon CARE Fertility will ease patients into jargon, where necessary, as they are introduced to the procedure that awaits them: consultation rooms, man and woman separated into gendered sample rooms, the testing, anaesthetics, the viewing onscreen of each action. Back stage, across the 'Red Line' that separates the non-medicalised, naturally lit, spacious consultation suite from the treatment rooms and labs, Beacon CARE staff describe the pathways that form the 'male' side of the lab to the 'female' side. The hyperfilters remove organic contaminants. Grade D filter is the minimum requirement. Beacon is a Grade C or even Grade B. Origio ScanLab can operate to Grade A air quality standard, New Brunswick incubator, controlled environment, positive pressure, high O2. Operators now also use a high sensitivity joystick to move the sperm into the egg. They refer to themselves as 'facilitators'. It can gather more sperm physically to inject straight into the egg. The embryo biopsies take place here also. It requires laser technology to penetrate the protective zona pellucid of the oocyte.

Staff may not use the term 'social freezing' when discussing cryopreservation of oocytes with a female patient, but some detail may need to be explained to males about what is required of them in the sample room; for others, what takes place remains unsaid. The woman, understandably, will require most information and reassurance about the invasive procedures ahead. It is not a straight-forward procedure but the Deirdres in every secondary school in the world need to know.

Back in time, in ND, Donal was one of the (quiet) jokers of the class but school has no real interest for him. He had dreams of being a journalist or working in television and is seriously looking at media courses at DCU. He identified with Ryan's use of PGD technology, perhaps even for sex selection. Students were very forthcoming about their private lives and people within their outside worlds who have had experienced health issues and tied it in to the genetic screening theme. The area had close links with DCU's outreach programmes. However, NDt was worried that the 'weak students' in her class would not relate to such content.

SD, South Dublin

SD was selected because it was a co-ed boarding school of over three hundred students in an affluent area of Dublin. More than that, the study group of fifth years was unique in having mostly non-national students from more financially well off families around the world. SD has three hundred students, one hundred of which are non-national. The international flavour of this class was exceptional even by this school's averages. In a class of fifteen, just one or two were Irish nationals. Students who we spoke to throughout the sessions came from Germany, South Africa, Belgium, Kuala Lumpur and Bahrain.

Della was German. She told me she was unsure what her dad did for a living and her mother was a 'house woman'. She was careful not to reveal too much in interviews and finds it difficult to self-analyse. Douglas from Kuala Lumpur described himself as 'moody'. I was interested to see how different his Catholic upbringing might be in comparison to many Irish boys of his age. His father works for a large multi-national company in the UK where his family now live temporarily while Douglas boards in SD. He is technically minded. His interest in biology stems from it being a 'very precise' subject to him, but our activities are challenging these common practices: '[Biology is] black and white...yet there's grey areas as well so it's quite... interesting... sort of'.

SDt had a background in neuroscience before turning to teaching, as well as also being from outside Ireland. SDt, in his early thirties, is from the UK. SD was also selected as it was one of the few schools to use the online platform which was intended to be part of data collection in the original pedagogical plan (see Chapter 3).

Students assemble at chapel first thing every morning for 'meditation'. 'You sing a psalm and a hymn and one of the prefects reads the lesson,' Della tells me. According to the prospectus, the chapel is at the centre of its existence, both metaphorically and geographically. SD identifies itself as having an ethos that is 'informed by the Christian values of tolerance, kindness and selflessness that are so important within a close-knit community' (SDt prospectus). The pastoral ethos is reflected in its emphasis on communal singing and choral participation.

NDC, North Dublin city centre

NDC is another well-funded 'oasis' school. This institution is all male, and might seem to have a hint of siege identity from the outside. The school is hidden away from a more socially disadvantaged part of the city, with large steel and glass doors that barricade itself from the outside world. The school presents a strong Catholic ethos to the outside. Derek (NDC1) was somewhat of a rebel. His attitude in the film discussion was at times confrontational, with bored expressions, as if bothered by the setup and with other people's opinions. In a follow-up interview after the video sessions, he said that he was grateful for the experience of education, and was willing to explore a complex range of issues relating to sex selection, character trait selection, and genetic determinism in RGTs. These were all issues that concerned him. Derek has a relatively large family, although this is the result of a parent's second marriage a distance from Dublin, where he spends a lot of time.

Des (NDC2) was a more reserved student, covered extensively in Murphy (2008). He took a neutral position on the PGD in film discussions. But away from his peers in interview, he began demonstrating support in favour if the embryo as entity:

NDC2: ....implanting for fertility? No I mean I wouldn't. That ... that wouldn't involve actually destroying an embryo would it?
PM: ...But in some cases they may have to produce a few embryos.
NDC2: OK.
PM: Now that's not in all of them but in some cases. And....
NDC2: Yeah yeah, Mm.
PM: ...what's left there are generally called, they're literally called, 'spare embryos' because some will take and some won't.

(Des, NDC2, interview)

This whole conversation left Des uncomfortable about embryos' chances of survival. In group discussions, Damien (NDC3) argued the most strongly on behalf of protecting the embryo but not for overtly religious reasons:

NDC3: Yeah. I think nature has a certain balance like ... you can see it like in all species. But when we start interfering with that it does kind of create a kind of lob-sided then. And it's like the right gene pool or whatever isn't there any longer. Nature put us here like, we can't really dictate then...what's going on.

(Damien, NDC3, interview)

For Damien (NDC3), nature has a balance and human intervention can upset this balance. Humanity is definitely part of nature, but just one species of animal that can get carried away with its own arrogance. Damien also offers a scientific explanation for not using PGD: the gene pool could be upset. This was a unique feature of Damien's responses. He would keep conversations loosely within the philosophical scientific realm even when attempts were made in interviews to guide him towards his own personal interests or school. He may have made this demarcation himself – discussions could wander into the arena of philosophy at times and this is how he saw the content of film discussions. However Damien would seem to have an interest in science that goes beyond science class. Biology is his main interest. Without prompting, he would identify a controversy about science versus religion in evolution. He categorically said that he was not influenced by anyone else on the issue of evolution. Again and again he would return to this subject.

Later I attempted to guide him towards talking about interests besides biology or philosophy. After he talks about his favourite films, he brings the subject yet again back to the philosophy of biology. Clearly Damien used this opportunity as a platform for talking about those philosophical issues of biology that interested him.

NDC, similar to ND, revealed my simplifications in the compromised choosing of identities in debate. I had selected interviewees on the basis of where they positioned themselves with RGTs in class debates – 'anti-', 'pro-', or 'neutral'. The three selected interviewees seemed to change position from the film discussions. Damien did not agree with interference with the embryo but not because of any 'essence' or personhood of the embryo. RGTs 'interfered with nature'. Des, relatively neutral in debates, did indeed see a strong need to think of beneficiaries of RGT treatment, but was resolutely against any impact this might have on the personhood and integrity of the embryo. Derek, a rationalist, and staunchly pro-technology during each statement in film discussions, produced a case in interviews for not using PGD for any reason but essential health ones. Peer pressure in the public forum might be used as an explanation for the partial shift in position here – he wanted to portray himself as detached, confident, cool – but the complexity of each of their views reveals other facets of 'performing' in a debate which can occur publicly and in private, as described in Chapters 5 and 6. My own judgement in selecting three interviewees from the pro-technology, neutral and pro-embryo/nature sides of the debate was justified in terms of investigating further into cultural assumptions and the creation of frames regarding RGTs. However what is most interesting – and has repercussions for survey research into these kinds of discursive opinions – is that participants used the film discussion and the interviews as separate *locales* with *separate perspectives.*

LH, Louth

LH was another Catholic ethos school, this one being all female. The building was a large Victorian school with many cavernous rooms. Louth is a border county and is the most northerly of the study areas. Unlike teachers at the three Dublin schools and South Kildare, the teacher LHt did not participate in sessions in any way. Also, like staff at ND, LHt was in perpetual motion, with just enough time for introductions before leaving me with the class. There was one other constraint which made the LH activities difficult to manage. TY is divided into two large classes of forty students each. As with the pilot, LHt was concerned about using just one of these classes for the study, preferring to include both if possible. This made for an overwhelming number of participants in discussion. At various points during the film discussion, arguments descended into unmanageable, chaotic shouting over other viewpoints of multiple participants. The pilot had a similarly large class however – although not quite so – and Presenter B's TY group also numbered about forty. Presenter B did not seem to have discipline problems. The nature of the film discussion activity – the structured physical movement and expression of views – leaves sessions open for emotive, *performed,* arguments. Presenter B may have controlled and 'worked' his audience much better than I did. Then again, there is a marked difference between presentation and debate. I noted however that responses to film statements from a larger class did not explore the film's themes in the same depth. The students were very enthusiastic. The pedagogical format here did not do their enthusiasm justice. However from the accounts given by the three interviewees, the themes from the discussion resonated with the students and they talked about them for some time afterwards.

At the time of the film discussions, two major RGT news stories broke. On the same day, scientists in Newcastle, UK announced that they had cloned a human embryo and South Korean scientists had announced the differentiating stem cell lines derived from a human embryo, the first time this had been done successfully.[3] In film discussions, I mention these types of major media events, and science news stories tied in with the film, *If...Cloning Could Cure Us*. However, very few claimed to have heard of either cloning breakthroughs.

Cacophonous (*kakacosmic*?) debates aside, this school was a good example of strong types of framing strategies both in open forums and in one-to-one conversations. This first interviewee was unique to the study, a rarity in the study. Lorraine (LH1) had strong convictions on the subject of RGTs. She *believed strongly* that destroying an embryo at any stage was wrong. She told me of her deep religious belief and she insisted this had nothing to do with her parents (who are less religious), but was informed by a faith she had found for herself. 'Well I *am* a Catholic,' she said in response to my initial questions about her views on the technology discussed in the classroom.

Laura (LH2) was selected for interview because she found herself undecided on many of the statements in the film discussion. Laura brought up an interesting genetic duality in her family that might, she believed, contribute to her and her sister's different approaches to life – her parents. Laura's father is 'good at art', her mother 'good with computers'. Laura herself has an interesting and varied list of hobbies but seems drawn towards arts and drama whereas her sister is currently doing biomedical science. Laura tells me these home worlds may influence her, but the ethos of the school certainly would not in terms of views on RGTs. She is critical of the Catholic Church's authoritarian position on RGTs.

Lisa (LH3), who was strongly in favour of using HESC research, appeared to be a good friend of Laura's. In discourse analysis discussed later, there were some shared schema constructs between the two. They had debates with other students after the film discussion. History is a big interest of Lisa's. We also had interesting discussions about make-over and plastic surgery shows on TV which she cannot resist, and feels a certain shame in this. 'I'm kind of girly,' she feels a needs to say, in justification. Lisa decided not to take biology, but she began liking it through the science elements of the Transition Year programme. Lisa, like the rest of the LH class, hadn't been involved with the bioscientist sessions.

LH opened up a particular structural element reported by many educational ethnographies, none more so than Bourdieu and Passeron (1977), about non-verbal and tacit modes of power. When I was leaving after the film discussion session, the principal reminded me – with cries of argument still ringing in my ears – that the ethos in the school does not lend itself to question abortion. This was in response to an earlier discussion about my work, rather than a warning. However she did not question me further about classroom activities. Although LHt was driving this, the principal was the first point of contact at the recruitment stages. She would have been informed of the subject matter at the pilot stages. However, the abortion arguments would not necessarily have been a detail we discussed – and it is possible neither did the principal and LHt. It could be argued that the official ethos advocated in literature and verbally from schools such as LH or NDC does not reflect the day-to-day decision-making processes of teachers and principals who see immediate benefit from allowing open discussions and challenging orthodoxy. Is it an example of passive enforcement, much like the Catholic Church's stance on IVF? Do these critical events that are anathema to the ethos occur regularly, slipping beneath the radar? Perhaps the principal gave me the benefit of the doubt.

SK, South Kildare

The SK village lies close to the Kildare/Wicklow border. It is a farming community and SKt, as he says himself, 'keeps a few sheep'. Our study group were Transition Years with third and fifth years brought in for one session. SK formed the strongest connection with the research project and, by extension, with the university. This was in no small part due to SKt's enthusiasm for using the university as a resource and his professed desire to allow a more meaningful way of teaching science enter the classroom than was heretofore available to his classes. He felt media, old and new, has a strong part to play in this. Ultimately though, each student is responsible for his or her own education. SKt's mission is one of encouragement for those with a glimmer of interest in science, to help them pursue this path and excel. He was born and educated in the SW town, which is not far geographically or socioeconomically from SK. He could have been in the army. His inspiration for taking up science was the influence of his old science teacher from SW. He called SK a 'mixed ability school, a vocational school-cum-community college'. He had been there for thirty years, teaching two generations of students, and building a community rapport. He is saddened that there is not 'a tradition of education' in the area.

Since being involved with the research programme, SKt has had two years of work placement for TY students in the NICB and was constantly approaching the university for Young Scientist Exhibition projects to assist his students. He insisted on an organised learning plan. He also wanted our project to have a very definite learning outcome, that was 'inquiry-based' and 'problem-solving'. He contributed to the development of Junior Certificate science and Leaving Certificate physics and chemistry.

SW, South Wicklow

SW is also a rural school, on the border of two counties. I attended this secondary school myself, growing up in this area. SWt, similar to LHt, did not participate in sessions. She talks enthusiastically about using discussion activities to talk about RGTs, however. Where they are raised, it is not in the context of a social, citizenship, or philosophical reading of identity, but wider Irish society only, such as the monoculture of Ireland compared with other countries. She is somewhat disappointed with the curriculum change. It is not because of the contemporary issues strand – she talks about encouraging students to explore AIDs, global warming, genetic testing, topics that do arise from time to time – but because it has become a 'more vague, less tangible biology for kids'. She picked enzymes, photosynthesis, transport, and osmosis as examples. She expected an exam question about the pros and cons of biotechnology soon. SWt, not unlike SDt, is concerned where RGTs might be leading us. She feels an international decision needs to be reached for guidance. With regard to the current curriculum, she wonders how could a topic like bioenergetics be relevant to today's young people, particularly those who are never going to be in the industry. It is 'hard to connect with them'. She herself feels far removed from biopharma and cancer research. As mentioned, SW is located relatively close to SK and has a similar rural setting. In some ways, it is even more isolated, as there is no hi-tech industry nearby. 'Manufacturing and industry is not big around here.' Agriculture is the main local economy. The community of rural Ireland around these parts have a word, *meitheal*, which means gathered around a common cause, perhaps Latour's community of concern.

SWt is motivated by working within the education system. It provides a good home-work balance, and is conducive to family. She finds SW very supportive of her. However, something is amiss. When we talk about classroom outsiders, there appears to be a few students each year that she cannot reach. But it is not just the number of 'weaker students' – this story is similar to ND and SK. It is something else. Although she praises her current final years for their enthusiasm, the fifth years (the participants of this study) are less engaged. The more SWt talks about this disengagement, the more it seems to be a general trend from first to final year:

> As they get older, they get less inclined for chit chat, more withdrawn. Something is happening here in the system...It might be a problem with our school. Pressure is put on them from third year on.
>
> (SWt, female, interview)

Presenter B told me after his session that SW stood out for him as the one school where the teacher said there was practically no interest in science among the students. This was reflected in classroom discussions, which were muted. At first I was disappointed that my old school should feel this lack of enthusiasm for RGTs, identity, and society. It impacted somewhat on my theoretical assumptions about an emerging life politics beyond the curriculum. However, close analysis of transcripts characterised SW as another type of reaction to our activities, not least one stifled by the pressures of senior cycle. The low-key dynamic of classrooms threw up phrases like 'of course he should do the best for his child' or 'of course you shouldn't select for sex or sporting abilities'. The tone implies that there could be no other option in each case, and that perhaps argument is futile. Could this 'common-sense' attitude – identified in other science education contexts by Solomon (1994) – a one-sided, rational approach, be peculiar to the locality?

The bioscientist presenters

The intention of the bioscientist sessions was to provide a tangible link to the outside world [vis-à-vis] third level biotechnology research. As framing was a subject of my original discourse analysis, it might be appropriate to say that this activity was framed to audiences as *a glimpse into the world of the scientist* or *bioscientists have a specialised job to do in the treatment of disease.* To participating teachers, the activity could also be framed as a career option for young people, and be aligned further with other activities in Ireland designed to entice young people into the sciences. For the objective of this project, however, it was more important that it was a discursive forum where young people might gain access to third level science and *challenge* these practitioners. It was an exercise in issue engagement with young people. In communications, business, and politics, people present their work to be received critically. For the presenters themselves, it was a communication exercise, having more importance in Nowotny *et al.*'s (2001) Mode-2 science and Giddens' (1991) expert systems. The researchers who gave their time to this project are referred to here as Presenter A and Presenter B. Recruitment was a simple process. They were the only respondents to an email at the time and were given the go-ahead by the Institute's director to participate. The format for the bioscientist sessions was developed during the pilot stages, partly from improvised initiatives put forward by Presenter A. The session followed a set structure: in the first twenty minutes, the bioscientist presented. They talked about their motivations, their day-to-day work as university entry points and academic paths. Presenter A often took a wider view than Presenter B, sometimes veering into ethical issues. The presenters were asked to focus on their own work, but to relate it as much as possible to their audience and to society. Both presenters used references to genetics in news media or science fiction. The next ten minutes were given over to a questions and answer session, which in most schools was quiet. The final part was a more informal focus group structure where I and the researcher moved around clusters of participants and talked about anything relating to RGTs, society, and careers. Four school groups were fifth years and two were Transition Years.

Each speaker had a different style. Presenter A, authoritative and demanding, got attention very easily with tales of his Australian adventures or his concerns for cancer treatment. He was inclined to link his work more to society and possible outcomes of research leading to cures. He could be preachy about avoiding cancer, including in his presentation a 'Good Living Guide'. In the early days, he appeared a little arrogant, which worked in his favour when getting student attention, and winning over teachers.

Presenter B was more relaxed, presenting in a soft 'cultured rural' accent. He preferred to focus on college life, speaking more about travel during work and experimental problems than other aspects of wider society. He would make no great claims about outcomes except to say he was a link in the chain and his main focus was getting a PhD. While Presenter A's 'day-job' was research assistant at the NICB working on computer modelling systems for genetic markers that might reveal the presence of cancer, Presenter B was reading for a PhD on the application of genetic markers for cancer and diabetes. Where Presenter A was the manager, Presenter B was the charmer. Presenter B played in a band and he had this experience to draw from in his talk. His general amiability and approach made him well liked in some schools. Presenter A was instrumental in making the session structure a success as this entry from my pilot fieldnotes demonstrates:

> Once [the teacher] left, Presenter A suggested the 'break-out' session as per yesterday at ND1. Here is where the remaining TY students finally relaxed and talked to us, so this could be a worthy data collection tool. The magic words were: 'Talk away yourselves'. The class divided into natural groups of friends, to which Presenter A circulated...In the car back to Dublin, Presenter A hoped that some of the class were inspired to take up science.
>
> (Pilot fieldnotes)

However another important factor in the success of this set of activities was the relationship between each presenter and their 'audiences'. Both were early twenty-somethings, not far ahead of their students in life. Connections were therefore developed from early on.

Summary

Esposito tells us in the (a)modern classic *Bíos* that by thinking and talking of whom and what should not exist beyond humanity we have entered:

> a limit point beyond which the entire biopolitical horizon risks entering in a lethal contradiction with itself. This doesn't mean that we can turn back the clock, perhaps reactivating the ancient figures of sovereign power. It isn't possible today to imagine a politics that doesn't turn to life as such, that doesn't look at the citizen from the point of view of his living body. But this can happen reciprocally in opposite forms that put into play the different meanings of biopolitics: on the one hand the self-destructive revolt of immunity against itself or the opening to its reversal in community.
>
> (Campbell in Esposito, 2008: xviii, quoting Esposito, 2002: 170)

The *munus* meaning present, having the same etymological root as community and *immunitas*, the immunitary impulse (Esposito, 2008), also has three meanings – two nouns (a gift, and to be in current time or *esse*) and a verb (to give, perform). What young people who see this view of the future are bearing witness to is the tendency to stay close to the authentic self (Giddens, 1991), and resist the gift-giving into community that leaves the self exposed. Further up in structure, the *immunitas* is continued as some form of community (school, town, state, human group) which also immunises itself against this tendency to a forced gift-giving. The liberal individualism of capitalism that facilitates greater choice of genes, babies, life-planning and futures themselves is the perfect immunisation.

And so with this chapter, argument continues – given the cultural/strategic restraints of education, society and governance at large, how can our young protagonists deal with hard borders at the edges of the classroom, when negotiating futures, perhaps pre-determined like Willis's (1981) 'lads'? The constructed boundaries of the science subject recede even further when we see the locales or fields of film, work, of school, Science, Nature and potential futures. This chapter looked at the embodied sites as identities, students, and how they might connect with something real to them in the biosciences, related to RGTs, these *sites* or the more abstract *locales*, that might include beliefs. The competing scripts that young people must 'read' and 'perform' makes it difficult for them in any class to negotiate their own sense of identity, but the boundaries of science are particularly difficult, as we analyse in more detail in Chapter 5. We will now look more closely at curriculum matters and see the boundaries put around biology.

Notes

1 This chapter is a development on the movie-screen-to-performed-action politics of young people in Murphy (2008).
2 Other programmes in that BBC TV series intended for Western audiences and classrooms included *If….The Oil Runs Out* and *If…We Stop Giving Aid to Africa (If… Cloning Could Cure Us*, 2004b).
3 An interesting sub-plot developed to this story when the South Korean team led by Woo Suk Hwang was charged with scientific fraud involving this breakthrough. Question marks were also raised over another scientific claim made by this team – Snuppy, the world's first cloned dog. Woo has since been suspended from Seoul National University.

Chapter 5

The *bios* in education

This chapter looks at curricular and pedagogical matters in more detail for biology education, building on the literature and school-based empirical work from previous chapters. Having presented experiences in the previous chapter of young people and the grounded sociology of bioscience education, I want to now critically engage with the potential interactions between the two STSs – Science, Technology Studies and science-technology-society in education, while also looking back a little on European biology curricula taking up from the cultural and socioscientific shaping of Chapter 2.

When we speak of 'science literacy', some science communication scholars suddenly begin to hear the deficit model, the assumption that a scientific expert will always perform a kind of pedagogy from the position of greater assumed knowledge. What I will focus on here is other transferrable skills and literacies – we might call them *representational literacies* brought about by argumentation (this is performative, the mediation part) and consideration of biopolitical dimensions (the conceptual witnessing part). Life politics escapes the well-researched notion of biopolitical control. So teachers can be encouraged to use discursive tools, particularly positioning and argumentation, in biology class. Consideration of 'Mode-2' (bio)science in education prepares budding scientists, and the rest of society, for a different conceptual way of perceiving science, or sciences and technologies, that now exist – as relevant, embedded in or with society and inherently political. Formal, informal and non-formal education can be integrated to do this.[1]

The central methodology that I use in classrooms is not new – I have to acknowledge the Wellcome Trust and their 'Genetic Futures' activities (Baldwin, 1999; Townsend, 1999). But what is offered here is the conceptual idea that this communicative, argumentative approach is the flip side of biology knowledge processes and its pedagogy. It does not exclude biological literacy, but one 'side' helps the other.

There are more subtle sociological and historical processes that impact on the curriculum, as we saw from European and Irish state exams and education systems of Chapter 2. How do we move forward, to a potential link to *res publica*, 'things that are made public'?

The biology curriculum

Haraway said of biology: 'Two different words that have the same form, biology has always meant the thing itself and knowledge of what it is' (Franklin, 2006: 168). According to Rose (2007) (and Canguilhem in 1966), the term 'biology' came into popular use (separately) in 1802, by Jean-Batiste Lamarck, Ludolph Treviranus and Lorenz Oken. It was a re-invention of vitalism according to Rose (ibid.). In Oken's case, his idealism was as influential on German intellectual thought as his *naturphilosophie*. As further elaborated by Fuller (2011), the growing Weissman divide between biology and sociology, the *cordon sanitaire*, echoes the widening gulf between disciplinary definitions of life and non-life (Rose, 2007: 270, n5, citing Keller, 2002). However, paradoxically, life processes became understood through a machine and communication code ontology, referred to as *organicism* by Haraway (1976) and Franklin (2006); for Deleuze and Guattari (1987) the 'machinic phylum' goes further, in the Latour-like way in which knowable atoms, molecules and global structures are 'phylogenetically' linked, organised and self-assembled in world biopolitics.

Biology students learn in the earliest chapters of biology books that 'life is' metabolism, responsiveness to the environment, ingestion, digestion. For STS philosophers, this split between the organicism of life processes and non-life strongly sequesters life politics in science, as was evident in the early development of science textbooks. Also ontology and political thought shored up the division. As faculties of science grew quickly across universities of Europe, this filtered down into secondary education. We already mentioned the influence of Spencer's (1864–1867) and Huxley's (1875) first biology books in the UK on curricula education. Barbera (*et al.*, 1999) state that they had a similar influence in Spain. Martínez-Gracia *et al.* (2006), in their comprehensive review of genetics content in secondary curricula in Spain find no mention of 'ethical' or 'social' issues, the signifiers of biosociality. In Italy the *liceo scientifico* schools have foundational mix in their inclusion of Latin, philosophy and history as well as science.

If we look at a standard textbook to support the Irish Leaving Certificate (Freeman, 2011), the first striking element in the introductory material is the patented skills required of biology students of 'mastering biology'. Even the preliminary material contains a photo of *Cynus atratus*, the black swan with references only to beak colouration hypothesis. Is this the opportunity to discuss the standard dilemma in the philosophy of science or popular culture? Biosociality emerges here, connected to life politics. It enters in the shape of Pasteur's experiment on cell spontaneous generation (Latour's study of pasteurisation has been an accepted STS alternative history of Pasteur). 'Phylogenies and History' is another obvious place for STS perspectives. Engagement with young people was an obvious reason for inclusion of 'what killed the dinosaurs?' section as part of this section (ibid. 490).

Section 51 of Freeman (ibid.) on behavioural ecology takes on an anthropomorphic set of questions, propositions as important to human teenagers as to the non-humans:

'What should I eat?' (ibid. 1021)
'Who should I mate with?' (ibid. 1022)
'Where should I live?' (ibid. 1026)
'How should I communicate?' (ibid. 1027)
'When should I cooperate?' (ibid. 1031)

When compared to textbooks in the 1980s, this effort towards student relevance does bring the subject closer to life politics and biosociality. However, communication in biological science, is defined as

> ...any process in which a signal from one individual modifies the behaviour of a recipient individual. A signal is any information-containing behaviour or characteristic. Communication is a crucial component of animal behaviour. It creates a stimulus that elicits a response.
>
> (Freeman, 2011: 1027)

This is Behaviourist-Skinnerian in orientation, and relates only to animals, it appears. 'By definition' it continues 'communication is a social process. For communication to occur, it is not enough that a signal is sent; the signal must be received and acted on' (ibid. 1027). There is an assumption here that we cannot know communication in biology without observing action. However the modes of communication are only divided into acoustic and visual (ibid. 1029). (Subatomic and ecosystem exchanges are not part of this social process.) Yet information is passed in DNA, and particularly messenger RNA. It may be information, not communication. But the observer at least is there, so a network created. And there is a local morality. The Animal Kingdom exists on 'reliable signals' (ibid. 1030). 'Honeybees are always honest', Freeman tells us (Freeman, 2011: 1030). 'Honeybee dances consistently "tell the truth"' (ibid. 1030). Freeman's text is quite engaging, and laced with humour. Elsewhere, while referring to the Central Dogma of contemporary genetics (the thought collective of information technologies: DNA as information storage, transcribed via messenger RNA carrier to translate into protein, the active cell machinery and building blocks for the organism), and the exceptions to this Dogma, 'The punchline? Crick's hypothesis is a central concept in biology but cells, viruses and researchers aren't dogmatic about it' (ibid. 28). Non-humans have some representation in this mix.

The Deleuzianism of the Leaving Certificate biology book's defining of communication as that which cannot exist in non-human nature without action, as Bennet contends, via Nietzsche and Bergson, is 'an action-oriented perception on which our survival depends' (Bennet, 2010: xiv). Bennet's communicative action can be taken in the context of how Latour would see Lovelock's Gaia (detached, not caring of humans except in strictest of animal nature sense) mixed with values for the common good of our planet. Perhaps however, Western curricula are moving, at different rates, in the right direction. With the tools of pedagogy discussed soon, there are just a few more steps before reaching a life politics curriculum.

Pedagogy: belief, believing and making belief

De Certeau (1984) distinguishes between the strategies of the strong, the experts who provide a specialised knowledge in which to place themselves and others, and the tactics of the 'weak', the 'other', those who do not use the dominant scientific discourse, but use ruses and wit, full of phrases with double-meaning and who are capable of manipulating and repurposing the practices of the oppressors. Indigenous peoples throughout history

> often used the laws, practices, and representations that were imposed on them by force or by fascination to ends other than those of their conquerors; they made something else out of them...They metaphorised the dominant order; they made it function in another register.
>
> (Ibid. 32)

In Irish postcolonial nation history, we might recognise such tactics. Cultural, media and consumerist practices will be reconstituted back into the living culture of local values. Scientific discourse uses strategy, but I do not want to address it here as an ideology; it goes further as an *episteme* that controls multiple discourses (Foucault, 1970). Those who defy scientific evidence presented to them, or oppose a technology that has been presented as having enormous biomedical value, may draw from a *counter-episteme* such as 'going against nature', using de Certeau's idea of tactics to resist the majority theme that is culturally salient.

Teachers need both strategy and tactics for dealing with the constraints imposed by curriculum, time, space, money and biopolitical body positioning, discursive constraints that separates biology teaching practices from other subjects and topics, and areas like media science and science fiction, the separation of 'school', 'real science' and 'fiction'. Each discipline that has emerged from scholastic tradition is a thought collective (Fleck, 1979[1935]) with a different language. The crux of the matter for science teachers, just as it is for scientists, is the immunised locale that attempts to keep ideology out of science. The traditional argument has been that ideology is the reason science has been misapplied. The strong STS programme has yet to materialise in education, a myth where science is pure, only sullied by bad methods and bad culture (Bloor, 1991).

How does it work for teachers of the arts, humanities and politics? They all function perfectly well and in many cases the vibrancy of pedagogy calls for less objectivity. In fact, the critical pedagogy of Paolo Freire and Henry Giroux displays for us how young people need to see, witness, what they might consider 'false consciousness', but while this phenomenon is thanopolitically fake or not is for another consideration; important here is that the student is called into action by a piece of teacher's rhetoric. To believe that ideology becomes cleansed by 'good science' is profoundly mistaken when 'science' never operates by itself in a vacuum. This is a dangerous idea, but we are not concerned with what young *believe* when it comes to the communicative, argumentative aspect of life politics in education. They are 'beliefs' as *dispositifs,* what De Certeau calls *strategy*; it is controlling, and Agamben (2009) translates from Foucault as *apparatus.* Ted Schatzki (Schatzki and Natter, 1996), a unique and influential practice theorist, draws on Foucault's *dispositifs*/apparatuses such as architecture (physical and conceptual spaces of geography), behaviours (non-discursive institutions) and discourses (propositions of legal, scientific, philosophical, moral, rule-types). As we saw in the previous chapter, humans are social bodies in this mix, history and structure scripted into their action. 'Belief in the part one is playing' is how Goffman (1959: 28) described it, sometimes deluding the audience. There is no exchange for *communitas*, just the thing of community stolen from the giver of gift, the equal in traditional understandings of political rhetoric. It is performed rhetoric, the act of pedagogy, the summoning of opinion.

In Campbell's (2011) reading of Heidegger's (1977) *Letter on Humanism*, Heidegger's famous typewriter, as opposed to the pen, introduces technology and somehow makes things unclean, unclear. The pen is the proper, also indicated by De Certeau (1984) as a move toward the subject pole, unlike the improper, the typewriter (op.cit.). Likewise, the public mass is clearly bounded (written), but places individuals into the fug. The masses, the publics, the openness is the ultimate objectification, as also analysed by Agamben (2000) and Esposito (2008). All of these readings of Heidegger perceive communication itself as a threat; once we have the language for life, for human life its origins and finitude, then life itself is controlled. The threat 'is both biological and political... individual beings who are open to communication become both subjects and objects of a power over life' (Campbell, 2011: 25). Within this totality, the improper use of language itself threatens the very 'essence of humanity' (ibid., 25, citing Heidegger (1977), *Letter on Humanism*, p198). We must also remember that Heidegger was of the opinion that essence in biology went beyond biology as a technology (Heidegger, 2001: 296).

For Latour, the power arises in strange places, and sometimes it is not present at all where you think it is. So domination from improper language is not at all certain – what has more worrying surety is how, according to Campbell, Heidegger in this text separates the proper human (the Platonic idea) from the improper man. And there, suddenly, appears the human Being whose life is not worth living, *homo sacer* (Agamben, 1998). The dangers in the communication of the biosciences to which new biopolitical philosophers are alerting us are significant for education.

By Campbell's (2011) analysis on proper life and improper life, again in the project of seeing the separation of *bíos* from *zőë*, and as we saw glimpses of earlier in this chapter, it is actually Heidegger's 'improper human' that appears in biology books, with flesh, blood and organs. Here there is no historicity of philosophy, or indeed religiosity, of biology. Also, as we said in Chapter 3, where is contemporary theory? Where is the use of media, a space of socialisation and politicisation for young people (Buckingham, 2003, citing Chaffee and Yang, 1990) as well as scientists and science communicators? The collisions of bio-art, eco-philosophy and bio-philosophy that are cutting edge in academia? Stephen Jay Gould famously devised Non-Overlapping Magisteria (NOMA), and it is this Weissman separation that constantly prevents the 'contamination' of ideology in science by its very objectivity.

What follows are some accounts of school negotiating with the boundaries of educational locale and 'something else' for students, but also, crucially, the importance of rhetoric for the teacher and how students trust the expert systems as Giddens (1991) defines them. Is the expert the teacher or is the 'the outsider'?

While visiting the six schools, NDt and SWt often raised the issue of time constraints on facilitating discussion activities in biology class themselves similar to what I was running. Many teachers see Transition Year as the natural home for a module of these types of activities. In Ireland, this is fifth year, the 'break' before senior cycle where, depending on the resources of the school, students had a bit freedom to choose interesting topics to study for that year. This wasn't part of the curriculum.

To teachers, pedagogical practices facilitate students to be *informed* of science and other viewpoints, to see the complexities and, if possible, draw conclusions. A right or wrong answer supplied by the teacher is not part of their understanding of this pedagogy. One religion teacher was cautious about over-emphasising or highlighting a pro- or anti-abortion ideology to her class, but requested to be given more science content to be in a position to answer queries for students about stem cell research. Objectivity was a particular concern for SKt. He was the most enthusiastic participating teacher. He required an organised learning plan. He also wanted this project with me (several weeks of the film sessions, debate and other combined pedagogical techniques which were not always successful) to have a very definite learning outcome.

Teachers felt they needed to be aware of their boundaries. SKt described the demarcations and interdisciplinary practices between, for example, science and religion:

> If you're doing say fertilisation; *in vitro* fertilisation is an interest [to students], contraception is an interest. If you are prepared to give that information. But now you also have to be conscious that you may be crossing over on somebody else's subject first of all, like religion. And number two then, you must be conscious that the information you are giving: is it accurate, is it up-to-date? And number three then, are you in a position to give them information on contraceptives and you know where [they] are available? And, you know, you're sort of, are always of the opinion like ….if I was doing, in particular, evolution, I would never take a stance on say the difference between spontaneous generation and the fact that… it involved down

along from one to the other. Y'know that you start off simple with a flash of lightning and the[n comes] protein and then you move down along.

(SKt, interview)

SKt appeared to be constantly drawing back from the controversy of these issues. Once he had defined that a border exists between the religion subject and the biology subject, he seemed to hesitate, even in recollection, from discourse on sexuality (as opposed to sexual reproduction), abortion, or availability of contraceptives. While he is concerned about crossing boundaries into religion, likewise a participating religious teacher had concerns about science, but significantly she would be willing to make that border crossing if she had the information. We will return to this religion/science class combination in the final chapter.

When I mention to the SD principal that research has shown (Levinson and Turner, 2001) that teachers in the UK were uncomfortable dealing with humanities issues – exam practice and training being big issues – he defended science teachers, stating that it may not be an unwillingness or inability on the part of science teachers as much as a reluctance to mix up different subjects, each with its own set of practices. He was a biology teacher and a drama teacher. He had used drama techniques in the past in science class but only in exceptional circumstances. The SD principal liked to keep them as separate disciplines. Teaching scientific thinking, he says is teaching a different way of thinking. Cross-disciplines require a great deal of teacher collaboration.

Back in SK, SKt explained to me how he would go about setting up a discussion on a controversial issue in class. First, he would start the scientific topic, 'setting the scene'. Then he would try to include as much practical lab work as possible under restrictive circumstances. He might then refer to recent news items that relate to this particular experiment or lesson. He gives me an example: the news reported a man who was caught involved in fraud in Germany using DNA profiling (there are many cases of fake genealogy tests, etc.). He says he might refer to religion to 'stir things up'. The scientific method is still his main focus, though. He says he will discuss anything but: 'Background info for some of these topics is non-existent in these students' lives. They get their science from *The Simpsons*, *CSI*… Children don't read newspapers.' This appears to be what Brossard and Schefeule (2013) also found.

SWt felt students are far removed from biopharma or cancer research, as there is no such research occurring in the area, so no connection for local students. To SKt, school science itself is removed from the experience of young people. Experiential knowledge only will cause young people to ask questions about media science in class. Students do not, he insists, get science from newspapers or current affairs coverage. That said, large-scale media coverage like Dolly or a natural disaster will 'capture imaginations'. SKt suggests that media and pedagogy may converge in art and science subjects and may be of particular benefit to fringe students, including NDCt and SDt.

SKt explains the constraints on RGT discourse in normal biology class, principally that of time:

> There are constraints in the classroom situation and…it might be something that only comes with experience as a teacher, that … you need to prompt the following area as a ten minuter to sort of get back, get people back on track.
> (SKt, interview)

Recall the 'reading the script' anecdote from SKt in Chapter 4. SKt feels particularly conscious about the inroads of these external ideas. I was an outsider here, so it was easy for me to invade, to colonise, without fully understanding the pressure the teachers were under. But we all rely on these expert systems – teachers, students, academics. In SKt's educational vision, choice is important. To him, students can and should select from his curricular content, but also the films they see, the books they read, the outsiders they come in contact with talking about their work. He used the terms 'goal-orientated' or 'problem-based learning', recent concepts in science education. One school in particular shared this vision of teaching practices, yet mounted a collective challenge, because this vision was lacking, as discussed in the final chapter.

Students' voiced feelings of general powerlessness took two forms: from a sociological perspective, the more serious type of powerlessness reflects their place in the world and inability to engage with issues that can be described as life politics – abortion, euthanasia, contraception, fashion. A second type is a more mundane structural type that is more acceptable in normative terms: the curriculum exists and there is nothing students can do about it. Many had a set understanding of biology as a school subject. While most wanted films in class all the time, a few appreciated media and discussion but wanted to get on with the science. Real science to them was not this discursive or performative.

The spatial arrangements in schools and the different interaction of students that discussions and presentations offered created a new pedagogic dynamic for the subject; the change of classrooms to view the film, or, for those schools that had a dedicated room, the location of the TV in the school. This change of geography and tools of work – new media replacing traditional media or labs – meant a change of language, a different way of dealing with science outside the lab than in a science classroom. They were going 'to see a film'. They were not 'doing science' (see Jiménez-Aleixandre (2000) for a similar quandary). Although every class had its sulkier elements whom did not anticipate this would be a *good* film (and there was justification; students giggled through some scenes, as we shall see), there was generally a light mood.

The opening scene in *The Gift* begins with a shower scene of 16-year-old boys after a tennis match, Mark and his friends, shown naked from the waist up. This, predictably, drew laughs and jeers from students. Interestingly, only one teacher expressed concerns about the partial nudity of this shower scene, deeming it inappropriate (this problem was not highlighted in the Wellcome Trust's evaluation

(Wellcome Trust, 2000). For NDCt parts of the film were 'cheesy', a comment that cropped up often with students. Having initially piloted *If...* (2004), I wanted an equal division of schools watching each film. However four of the six ended up viewing *The Gift*, as I felt it had an appropriate exposition of ideas for identity-society decision-making, and the most adequate for debate. Despite her critique, NDCt was willing to use it anyway based on how it explored the issues, rather than its style. A teacher on work placement at the National Centre for Cellular Biotechnology which funded my research showed less interest showing a film than other activities such as the biotechnologist visit, a talk on ethics, or some resource graphics, and text of the lab materials, suitable for PowerPoint presentations.

One session took place on Darwin Day. There was an oratorical style from the teacher, soundbytes, and rhetoric. In front of me, noisier individuals were challenged (the internal outsiders identified by these students themselves and their classmates) to come up with answers to the teacher's agitated one-word questions. 'Genetics?' The teacher appeared to be playing word association with the class. But there was silence. No response. 'Media?' Still silence. 'You there. Yourself? Interested?' To describe the contingency of science, Darwin was used as an example: 'Darwin had technical ability and knowledge enough to make choices. Natural selection was not part of his plan'. Yet introducing Darwin suddenly brought forth an unlikely scenario of power relations over a group of disinterested students. The teacher then said: 'Today is Darwin's birthday'. A student replies, 'It's mine too'. It may not have been important to our study whether or not it was indeed the student's birthday. Perhaps the student often responds sarcastically, in front of peers. What was important was that a potential connection was met with this short response from the teacher: 'Good man'. The teacher's response also appeared sarcastic. Later in this open discussion, the teacher asked the class what they knew about mutations. The following field entry recalls the event:

> [Teacher] talks about mutations. 'Like *X-Men*,' [Outsider student] shouts up. Laughter. [Teacher] is rattled. [S/h]e wants [the student] to contribute. 'Explain yourself,' [s/he] says. [Teacher] genuinely wants him to engage... [Student] says '*X-Men* were funny looking but had special powers and no-one liked them.' Then: 'They stuck together and became superheroes. It's cool cos they killed people and blew things up.' He plays to the gallery – and it works – before withdrawing into a place he seems to visit sometimes – careful, tentative, unconfident. [Teacher] demands more answers but the burrow is being dug deeper. 'I really want to know,' says [teacher]. 'I need to know what's in that head of yours. Look, you and I are in different worlds. I don't know anything about yours and you don't know anything about mine.'
>
> (Fieldnotes, SK newspaper analysis session)

In many ways, the teacher and I had similar intentions. However, I wanted to see how students deal with the life politics of RGTs that is happening 'on the outside', through discussion, debate, rhetoric, lifting out of the inner world of a cleaned-up science classroom. The teacher's perceived goal is to get them to access the 'true science' from the outside, thereby enriching their lives. I needed to balance teacher requirements with my own concern that I may be leaving many participants behind. Both teacher and student used rhetoric on different levels, and neither attempt at communication was successful. The outsiders however did get some kudos from the gallery of peers.

How science studies STS can inform science education STS

Although they are quite frequently seen as synonyms, there is a disciplinary difference between STS as used in science education (usually an acronym for science-technology-society) and Science and Technology Studies, an interdisciplinary field that draws from the history, philosophy and sociology of science. Science education has evolved into many research areas, although by no means silos, such as socioscientific issues (SSI), science controversy, the nature of science (NoS), naive science, science-in-context, and many others. Argumentation features in science education literature (Chowning *et al.*, 2012; Driver *et al.*, 2000; Jiménez-Aleixandre *et al.*, 2000; Osborne *et al.*, 2004; Parker *et al.*, 1998; Sadler and Fowler, 2006; Sadler and Fowler, 2008; Simon *et al.*, 2006; Simmonneaux, 2001) as part of learning science *per se* but also for a wider scientific citizenship. Irwin (2006) and Horst (2008) use similar wider public pedagogical concepts in argumentation in the science communication literature, just as Latour, Fuller, Giddens, Habermas and many others dip constantly into the worlds of debated ideas.

As stated from the beginning, what I mean by pedagogy is the study of teaching and learning but also a wider 'public sphere' idea of sharing and co-constructing knowledge *a la* Habermas (2003), in several spheres as with Sloderdijk (2011). Learning theory combined with the latest science communication 'PEST' methodologies. We need to be aware not just of the opportunities to engage and be engaging, a competitive challenge in these mediated times, but also implement best practice learning. The repetition and follow-the-crowd of teenage practice and wider social practices needs also to be challenged. Acknowledging practices, while getting students to stop, think and argue for themselves. Even they have not yet stopped to think yet because, one hopes, they have yet to deal with a problem pregnancy where technology is the answer; the embodied, performed debate of the classroom floor about RGTs spurs on a thinking of a kind.

The study of how biology education represents ideas from the social, cultural, ethics and citizenship spheres crosses literature strands from science communication, biology education (formal and 'informal' learning methods) and the philosophy of science. Social issues in the domain of science education began with a number of projects in the 1970s, motivated in part by emerging fields of study at universities in North America and the UK. These became known as 'Science, Technology and Society (STS)' in the US and Canada and 'Science in a Social Context' in the UK (Aikenhead, 2003).

During the 1980s and 1990s, a number of STS-like school science curricula evolved, using titles such as 'public understanding of science' or 'citizen science', similar to what was going on in the sub-field of science communication (Aikenhead, 2003). In Canada, the Science-Technology-Society-Environment (STSE) movement emphasises a cross-disciplinary approach to science teaching based on student experienced in life and their environment.

Aikenhead (1988) clarified the expanding repertoire of science teaching strategies that a teacher could develop while implementing new science courses. Suggestions on how to use such strategies were also included. Ros Driver was a primary force investigating students' perceptions of the nature of science, not just the way they learned it, using her key content/process/social enterprise categories for science education (Driver *et al.*, 1996). Each of Driver's categories is embedded somewhere in a typical Wellcome Trust-produced lesson, developed by Baldwin (1999) and Townsend (1999), used in debates sessions in biology class reported in this book. STEM Futures, the Nuffield Curriculum Centre innovation, is an engineering example of this class-to-big-idea-in-society approach. Instead of closed loop systems of resilience in environmental technologies however, the focus here is on performative procedures embedded in RGTs, to ensure we do what is for the common good in human, and further-than-human, enhancement.

Solomon (1999) found curriculum innovation in science was no longer a technical matter for science education experts or teachers alone, but a complete cultural change was needed, a fresher, integrated way of looking at science subjects. Both STS (Aikenhead, 2003) and SPU (Osborne *et al.*, 2002) identify three key areas of curriculum development: *intended* curriculum (government policy); *translated* curriculum (textbooks and teachers' ideas about what will be taught); and *learned* curriculum (the concepts, capabilities, and attitudes that students actually take away with them). The introduction of bioethics to classrooms has been well documented in Asia. Darryl Macer (2000), Peter Taylor and Vaille Dawson have assessed the implementation of such programmes to facilitate moral dilemmas. Large-scale surveys such as ROSE, PISA, and expert reviews (Matterson and Holman, 2012) demonstrate an ambivalence (Michael, 1996) in how young people construct their world and knowledge systems around science.

In the Institute of Education's *Valuable Lessons* report (Levinson and Turner, 2001), teachers from one thousand secondary schools from the social and physical sciences highlighted the need for formal assessment of socioscientific issues based on 'well-reasoned argument'. Emphasis was put on building self-confidence, developing critical thinking and enabling students to deal with socioscientific issues in a balanced way, and engendering sensitivity towards the rights and needs of others. Humanities teachers were most confident when covering general ethical and social issues, but less confident about addressing socioscientific issues, considering the scientific facts incidental to their teaching of issues-based topics. Science educators felt that disregarding the science and its accuracy might diminish the subject. The report recommended that teachers of science should be supported to engage their students in discussion about these issues.

In a follow-up conference report (ibid.), science teachers highlighted the existence of formal assessment as a major factor in determining the level of coverage of socioscientific issues. Most science examinations reward knowledge and understanding rather than well-reasoned argument. In fact, in these early studies of socioscientific issues, RGTs were seen as a good test case. In a study of 14–15 year olds' attitudes to genetic screening for cystic fibrosis, Leach *et al.* (1996), identified argumentation skills rather than 'genetic knowledge' as the limiting factor where justification was weak in classroom arguments.

A key aspect of science in context is the use of narrative to explain both the science and the environment within which the science is described. *Breaking the Mould* (Osborne *et al.*, 2002) divided learning paths of a programme into 'Explanatory Stories' where scientific data and behaviour was represented using narrative devices, and 'Ideas-about-Science' where the 'big' concepts and issues created a forum to elaborate on science in the everyday. This informed the setting up of the SPU course, which developed into what is now 21st Century Science.

What are two particularly distinguishable items about the argumentation side of education STS is that the approach of Driver, Osborne and others has been to maintain the borders of science 'fact', making students think scientifically by the standards, and thought communities, that have been established and guiding them through conceptual change (Cobern, 1996). These are worthy goals for all students to do at some time in the education system. What the proposals here run up against is this very notion of arguing. Regarding argument, specific esoteric science will only be so much use; in arguing students are readying their positions, they are *pre*-positioning before propositioning. They have a lot of embodiment to get through. A strong rhetorical argument alone will not appeal to the Moderns, any more than it did for J. Craig Venter when he first presented in the boardrooms of venture capitalists.

However, it is through these young people that the history of biosciences can be directed and engaged with reflexively, as well as the future directions of RGTs. They may be here in biology class to train for a future scientific career, the objective of the STEM collective, but most of all they should be here to learn critically. The argumentation of objective theories of the sciences moves, and is positioned, within a broader arc of critical knowledge. Although rare, there have been important attempts to make connections between STS (science studies) and STS (science education) including Lemke (1990; 2001) and Wolff-Michael Roth. Lemke (1990) draws from the same well of pedagogy as I do here. In science studies, only Clayton Pierce (2013) has got there before me to write an STS study of how these concepts impact on the education system.

Debate in a life political curriculum

We are starting to see now how future governance can come into the classroom, for climate change, over-population, inter-generational politics, and transgenerational thinking. Perhaps even heading to space. But we need to zoom down again to smaller performances. A courtroom scene in Dayton, Tennessee, 1925. Another bio-orator, this time a crusading one, William Jennings Bryan sits after his brilliant *magna eloquium* before the local defending counsel Clarence Darrow, where John Scopes, a local teacher, stands in the dock accused of teaching evolution, which is against God's teaching. This is naturalistic science, epistemology, local politics and the court of public opinion. This is a fear from bible literalists of twentieth-century change, but maybe also a fear of cleansing, an *immunitas* driven by science? Nevertheless, the Scopes trial was a huge media affair in the US, a battle between the Moderns and the God-botherers.

Fast forward eighty years, and another case opens, this time in Pennsylvania. As a National Centre for Science Education report stated afterwards, using a rare STS flourish of networked things and concepts:

> *Kitzmiller v Dover* is now famous as the first test case on the constitutionality of teaching 'intelligent design' (ID) in public schools, involving a six-week trial in Harrisburg, Pennsylvania, dozens of lawyers and witnesses, nine expert witnesses, 342 filed legal documents, 400 exhibits, national and international media, subpoenas, depositions, lies, videotape, bacterial flagella, the Constitution, civil rights, education, science, religion, history, evolution, the meaning of life, divine intervention, and one recently appointed federal judge.
>
> (Matzke, 2006: 37)

And who should appear as an expert witness for Kitzmiller? Step forward, Professor Steve Fuller. Yes, our STS contrarian who pops up in every chapter here does it this time in style, representing Kitzmiller's defence in *Kitzmiller v* the 2004 version of the Stopes trial in reverse, where the teacher, Tammy Kitzmiller, brought her school to court for forcing Intelligent Design on the curriculum. Fuller testified that Dover had every right to do so. And so here we have yet another bio-orator to add to our list in Chapter 2. Fuller's reasoning was, as he answered the defence lawyer's questions, that science since Newton has often been reasoned from a monotheistic perspective, with a design impulse that is manifested from lab techniques to bioinformatics giving a God's eye view, and that scientists and those associated secondary scientific actors use persuasion to bring about this knowledge (TalkOrigins.net, 2005). Then in cross-examination, he went through his reasons why he thought Intelligent Design was a science, identifying order in nature that was separate from necessity or chance. The crux of ID's version of science is irreducible complexity, a point that reaches the God-of-the-gaps. Mendel's work on the hereditary nature of peas suddenly becomes infused with design, but not having (or demanding) the *naturalistic* justification – 'how would God do this?' – that Darwin's revolutionary rhetoric did, and brought to full design fruition in the twentieth century with chemical synthesis of DNA. This was shocking stuff to most in the courtroom and those reading the press unaccustomed to STS. It smacked, for many scientists, of postmodern rubbish, and Fuller was seen as an AWOL troll dragged up from the Science Wars[2]. Justification, as Boltanski and Thevenot (2006) have powerfully argued elsewhere, is the way of ordering by socialisation arguments, in different regimes (Sloterdijk would say 'spheres') of action in everyday life. Fuller is a New Romantic in a way, perhaps not the kind that dresses in Eighties fashion, but bringing along the spirit of Rousseau, Teuillhard de Chardin and the Jesuits. His actions angered many educators, as he recounted to the court, in a sometimes amusingly rambling and rapid style that caused the presiding judge to slow him down on occasion, that Intelligent Design had an intellectual history going back to Newton. Was he just being a trickster, a rogue, using a *ruse* of pedagogy instead of the *strategy* of one, as described earlier in De Certeau's (1984) idea of structure versus action? What was frustrating about Fuller's testimony was the visible support his testimony represents for religious-agenda teaching while a dominant scientific theory was at risk within Kitzmiller's education system, even if intellectually, rhetoric is his game. What was admirable about his stance in a way was that very rhetorical fight for an idea he posits is worth expressing; the battle from outside that needs to be mediated, much like climate change denialism. He brought, briefly, STS out in the open. He was hardly, after all, representing neo-Nazis or terrorists. Suddenly, to invoke a problematic phrase, the eyes of the world were gazing down on Kitzmiller. In reality, small town classrooms everywhere are affected as Gods and Monsters fight biopolitical battles overhead. The classroom had entered mediated politics in a major way.

There have been occasions in recent years when concerns have been raised in Britain and Europe about the spread of creation in classrooms. We must give students, and their educators, some credit. Toulmin's (1958) *Uses of Argument* is often quoted as an exemplar of how students can be equipped with practical reasoning skills using argument (Simmonneaux, 2001 and Jimenéz-Aleixandre *et al.*, 2000, both citing Toulmin, 1958), as is role-play, for example the Democs board games. Let the bio-orators disseminate old-school, and the students can *position* rather than role-play (Davies and Harre, 1997). Let them 'play' by framing around their position, as we see in the following chapters. The great liberatory educator Paolo Freire may not have included God in his lesson plan, but he would have allowed religious debate. Pedagogical theories should be combined with the latest science communication methodologies of science communication and STS in biology as should cutting edge, contemporary theory, bioart, ecophilosophy and bio-philosophy. Driver *et al.*'s typology (2000) – science content, science process, science as a social enterprise – is particularly instructive in this regard.

Notes

1 www.infed.org defines the three types of education thus: *formal education* is defined as the structured, graded system guided by assessment and leading consisting of primary, secondary and tertiary and eventually leading to full-time technical and professional

training; *informal education* is the lifelong process of acquisition of skills, values, knowledge of daily experience globally and locally including family, peers and media; and *non-formal* education is any organised educational activity outside the formal system, which may include community-based foundational courses and NGOs advocacy groups. All three can be considered part of biology pedagogy as described here (Smith, 2001).

2 For a good account of this curious academic bifurcation in the 1990s, see Nelkin (1996).

Chapter 6

'Political' science and framing

An education from media

Through mediation, we move closer in our analysis to what might be done Politically (capital 'P') by schools and students for RGTs and an emerging life politics. Chapter 6 looks at different forms of mediated action. Specifically, I examine more closely the concepts of *framing* and *positioning*. Framing is a type of analysis used in understanding how media represent content, and in turn how young people make sense of developments in the biosciences as they are presented in various forms of media. The positioning analysis maps where actors are in relation to an issue, if part of an interactive, face-to-face debate, or media event. Positioning means in relation to what they see (witness), how they engage with the mediated material, and with each other. It can be mediation from screen to body and from body to body. There are descriptions and quotes from young people as they physically and conceptually move around in debates, and how one argumentative person can influence the others, with or without back-up evidence or validated opinion.

Many science teachers across the world in high schools and secondary schools use press cuttings in their lessons. As we recount the 'outside' science as it were – those images and ideas that are extra-curricular – all media are crucial complements to formal bioscience education, not just the consequences of informal channels. Were science education to more broadly embrace all types of media in their repertoire, the inherently political would be made visible in the biosciences. This would be, for young people and teachers themselves, a co-construction of knowledge, an education from the myriad media sources of potential bioscientific fact/fiction partial objects on RGTs and other biosciences, as Latour would call them. These partial objects, fractured jigsaw-things, would be literal hybrids in the case of embryo-chimeras in the news, or *factishes* (the Latourian term for partial objects of fact; see also Disch, 2010) requiring more information and contestation in the forms of synthetic life like *Mycoplasma laboratorium*, nano-battery cells or even self-assemblers. What I also mean by 'an education' here is that there are many tools from media and cultural studies research of benefit to those working in science education research. The issue of framing has its own issues in media theory however, and we will consider this also.

This chapter then looks as much at how we (science communicators, STS scholars, educators and researchers) learn from media studies as how young people can learn from media. Again I will be taking particular examples from my PhD research to demonstrate immanent action of young people responding to the externalised biopolitical discourses of RGTs and wider biotechnology.

'Framing' 'science'

Bruno Latour's consistent critique of those he considers 'the Moderns' concerns the central issue of easy divisions of science versus non-science. The enterprise of empirical science, mentioned many times here, has its roots in the Natural Philosophy debates of Boyle and Hobbes (Shapin and Schaffer, 1985), and it was from here that a split developed that allowed Western modernists to 'frame' 'science'. Boyle's innovations have been, as we have said, to employ the mediation and witnessing that a science of any kind entails. When we look at the films at the disposal of youth today like *Gattaca*, *Splice* and indeed *Hulk*, *the Island*, *Iron Man*, and of course *CSI*, we see visions of science that can never be separated from thought collectives. Even documentaries have aesthetic representations of genes, mathematics and the quantum world. As many have commented, this is a public opening, a way for political and identity-construction that places visions of the sociotechnical at centre stage, particularly RGTs (Jensen, 2009; Michael *et al.*, 1997). The thought of Hollywood, Celera press relations, the *New York Times* and the operating procedures of a small biotech enterprise being part of the public pedagogy of genetic compositions sits easily with both Giddens' and Latour's descriptions of amodern assemblages. Latour laid out the common ground between science and media at the Dublin symposium, when he stated that scientists, in common with journalists – must 'count again....reweigh numbers of authority' (Latour, 2012).

News media are not generally majorly consumed by young people (Casero-Ripollés, 2010), and the diminishing of traditional formats in favour of devices and online platforms, and indeed free MOOCs, mean filtered news and public affairs items are a major sources of informal (e)learning. The use of 'old-style' clippings and healthy amounts of media analysis in the science classroom is invaluable for understanding agenda setting and storytelling for Generation TED science in the classroom, whether conflict of interest and 'bias' or difference of opinion or inaccuracies. Jarman and McClune (2002) found in their Newspapers in Science Education (NISE) project that, although teachers in Northern Ireland used newspapers as auxiliary teaching resources in science, few used them as a defined and central part of the lesson plan. As the audience and action research has shown for years, there is an elitism to the belief that young people must read the same newspapers as university researchers, or any newspaper, the plethora of studies that demonstrate the move online and to on-the-go devices demonstrates this. There are huge differences of understandings between various modes of existence (Latour, 2013) and therefore between regimes of justification (Boltanski and

Thevenot, 1991). Zimmerman *et al.* (2001) analysed and compared the content of news briefs on scientific research from publications in supermarkets and convenience stores and advice given by experts about how to read media reports about science critically. They then studied university students' requests for information as they evaluated brief reports about research. They found that the students' requests were less sophisticated than the advice given by science communicators about what to look for. Also however, there was a gap between expert advice and the level of detail found in articles about science process. Another finding was the difference between the scientific community's understanding of the term 'literacy' and the accepted one outside of the community.

In news media, there have been many examples of scare stories and moral panics over the last decade including, as we have covered, designer babies and saviour siblings (Challand, 2003) but also 'gene for' simplifications (Battles, 2002), frozen embryos (Cassidy, 2006), children with multiple genetic parents (Utton, 2004), chimera 'cow-people' (Martin and Caldwell, 2011), and of course abortion itself, particularly in Catholic countries. These are largely sequestered from education (Downes, 2005). Turney and Hornig-Priest have covered this moral panic in the media quite well for health and plant biotechnologies. This is risk perception and control, and editorial decision-making may commission such stories based on novelty and uncontrollable nature/human monster anxieties. Other news values include timeliness, and, as we shall see proximity. How do we see our future in terms of future people, generations and resources (material, environmental, human)? We are spanning governmentality to governance, and 'public opinion'; as we explored in Chapter 3, we question what 'attitudes' to science are, what investigating them means – the new adventures in opinion and knowledge for discussion and argument in classrooms and for wider social debates.

Frame analysis is one such, albeit fuzzy, methodology. The broad agreement about framing is that it constitutes implicit or deliberate acts of omission and emphases in text (Entman, 1993) (where I use 'text' broadly). I fit them in here among Latourian factishes/partial objects. And yet – despite this hybridization, the rhetoric of science is difficult to change.

Frames, framings and positioning in media debates

In this section I want to engage with the problematic of 'frame analysis' research, and how conversations and formal positioning of young people on RGTs are based on conversational resource strategies (Gamson, 1992) – media discourse, experiential knowledge, popular wisdom or a combination of these sources. The next chapter offers the outcome of this analysis. However here I will reflect on the realisation that frame analysis by itself, as understood by media researchers, is not enough to analyse the type of methodology employed in finding these framing ideas, these futures. They need to be heavily supplemented by a form of positioning analysis, the use of embodied action in sociolingustic discourse studies, and of course the STS of those on the socio-political spectrum, which is a rare combination in itself. I will start by looking closer at framing and frame analysis, then at the embodied nature of framing in real action, and how young people, like any of us, respond to proximal objects, to occasions, the fleeting form of culture 'out there', when the purpose of an activity may be deliberation and rational decision-making. I will draw attention to this mediated response by comparing classroom positions with a TV debate between a theologian, a bioscientist and a social affairs journalist.

William Gamson's (1992) *Talking Politics* is a particular, pragmatic influence on how I go about tracking this type of discursive action between the growing highways of information 'out there' and how it relates to, and is in its own way is changed by, all of us as listeners, viewers, readers, and ultimately doers of framing action ourselves. This is all the more important where the media consumer is now also the creator. Gamson's simple approach is to keep somewhat structural about this – the images and ideas, the symbols, the shored-up wisdom, the frames, are ultimately caught like rockpools from the ocean of noise surrounding us in simple conversations. The simple question is: how do we differentiate for opinion, and then make a case ourselves? Gamson's work draws on Goffman's performance metaphors in his structurational account of how agency and structure have a symbiotic relationship. People actively framing a concept deliver their 'lines' in a structural 'front stage' of society, while privately expressing other views 'back stage'. Goffman's (1974) *Frame Analysis* was the precursor to several types of discourse analyses across the disciplines. Potentially, the most basic framing that occurs in RGT discourse comes from actors in the abortion debate. News media-constructed frames may be assembled in a basic reductive polarisation of ideas (O'Mahony and Schäfer, 2005), questions, and answers. In their very name, the 'pro-life' frame creates an identity for the opponent as 'anti-life' whereas the 'pro-choice' frame creates the opponent's identity as an illiberal that frowns on individual freedom of choice. In relation to S&T discourse on such controversial matters, scientists sometimes opine that scientific facts get distorted in media packaging (Nisbet and Mooney, 2007). However, to be 'pro-life' or 'pro-choice' is to position oneself. Any amount of framings can still be used.

When we look at some definitions of a frame, two well-worn ones stand out at this stage, for example: 'Frames are principles of selection, emphasis and presentation composed of little tacit theories about what exists, what happens, and what matters' (Gitlin, 1980: 6), or:

> To frame is to select some aspects of a perceived reality and make them more salient in a communicating context, in such a way as to promote a particular problem definition, causal interpretation, moral evaluation, and/or treatment recommendation for the item described.
>
> (Entman, 1993: 52)

If we bring Latour riding in here, we see in frames Tarde's monads, the mediated, simpler, reduced version supersedes complexity (Latour *et al.*, 2012). Disch

(2010) also speaks of simplification in political representation, and we know there is also simplification of emotions and (hard) positions in media representation.

Frames need not be intentional or strategic. There can be a passivity about them. Where there is interactive social grouping in 'live' discourse or debate, young people may similarly frame biotechnology and RGTs in ways that are relevant to them, as empirical chapters here demonstrate, or engage with issues by calling on media resources.

In Giddens' (1991) late modernity, each individual person makes moral choices based on new globalised understandings of self-identity, and the media has a large role to play in this identity. Panel experts, chat shows, and celebrity life stories compete with their own personalised guidelines to lead a 'good life'. A society of interviews, confessions, and issue expositions has emerged. According to O'Mahony and Schäfer (2005), issue cultures are the media and communication processes by which a particular concern is highlighted and elevated in discourse through popular culture and media coverage. O'Mahony and Schäfer (2005: 112) refer to issue cultures as 'public beliefs, issue cycles, journalists' and media orientations, the interests and resources of societal actors, wider processes of inter-media and inter-textual knowledge production'. They are the media codes relating to other deep cultural tropes of humanity and nature that are caught up within practices of media circulation, news episode cycles, public salience, political and commercial lobbying, and intercultural references, studied in social movements' use of media.

In the issue culture process, an issue *package*, that is, how a piece is constructed for a media context, fundamentally depends on a *storyline* 'incorporating new events into its interpretive frames' (Gamson and Modigliani, 1989: 4), and sustained by deep cultural resonance and media practices. Negative news reports will have greater resonance if that storyline has salience, meaning it contains reference to issues close to a person's life: the familiar, the convenient, the relevant, or what Gamson (1992) calls the *proximate*[1]. In *Talking Politics*, Gamson (1992) says that certain tropes resonate with a distinct heritage of cultural myths, folktales, and half-remembered facts. In his work, he engaged with adult focus groups about media issues such as nuclear power and found respondents used certain references and phrases which had more cultural resonances than others.[2] Gamson suggests the existence of *cultural themes* that are 'safe, conventional, and normative' (ibid. 135), taken-for-granted linkages to the wider cultural world and, in common discourse that might be considered political, there are also *counterthemes* which are 'adversarial, contentious, oppositional' (ibid. 135). An example of a dominant theme is one conceptualising science as *progress*, whereas a subordinate countertheme may emerge from an environmentalist argument, *harmony with nature*, which may suggest that science is destructive. An argument for foreign policy may take on a *local state* theme or a *global responsibility* countertheme. RGT issues are examples of dichotomies with cultural resonances and many media arguments appeal to these: the heroic scientist versus evil clones or chimerical monsters running amok; the empowered individual versus the sense of identity corroded. Genetic science and technological innovation can be affixed to a cultural theme to demonstrate conflict, thereby gaining salience for those who empathise with a particular side – is it progress or is it destruction? Polarisation may be deemed necessary for argumentative clarity and a degree of salience. However, as will be shown in Chapter 7, this theme/countertheme complex in discourse may not be entirely incompatible, a phenomenon not explored by Gamson, but is in Snow *et al.* (1986). There are contemporary examples; current ecological framing may take on scientific *progress* themes while also using frames about *harmony with nature*.

Eder (1996) elaborates further on Gamson in relation to his work on how frames are grounded:

> We use and apply frames in order to sort the world, thus reducing the continuous stream of events to a limited number of significant events. Frames give to these selected events an objective meaning, thus disregarding subjective differences and idiosyncrasies ascribed to individual persons.
>
> (Eder, 1996: 166)

Besides their mix of polarity and valence, Jasanoff's reading of frames as non-deterministic suggests that they take on a life of their own in public discourse, leading to consequences beyond the expectations of those framing in media contexts (Jasanoff, 2005, citing Goffman, 1974). Whatever about public discourse, they seem to have taken on a life of their own across the academy. The lack of coherence on the subject is well documented (Entman, 1993; Kohring and Matthes, 2002). The context for frames in Gitlin (1980), Gamson and Modigliani (1988), and Entman (1993) is the deliberate framing in news media or motivational framing in social movements. What was once known as mass media may frame events as starting points for deliberation, such as the cloning of Dolly (Jasanoff, 2005: 24). These types of frames are actively constructed to persuade (ibid.). Snow and Benford (Koenig, 2004, citing Snow and Benford, 1988) however maintain that Goffman's original framing concept described a more subtle process, a 'conceptual scaffolding' pervasive in localised ways through all social (and in late modernity, mediated) interaction and thus less easy to control.

'Frames' as defined here draw heavily from Gamson (1992), collections of ideas around which an issue swirls in the actors' quest to negotiate meaning when making moral decisions, based on tacit knowledge of the past. But I prefer to use 'framings' and their future orientation, to what can or should be done, or what can happen. Frames may not always use reasoned argument. This is important. Gamson sees a particular issue of concern as a 'forest' (ibid. 117) through which others have already walked and left behind signposts, the media in particular. For many people, he says, these signposts are glanced at occasionally or ignored. People carry their own maps, compasses and ways of guiding themselves through the forest. Any issue can be a 'pocketbook' one, taken and stored for future justification in another context. Although conversations with local, proximate issues engaged Gamson's focus group more, faraway issues will also engage (ibid.), an effect that might be considered reasonable in an age of globalisation and mediation. A global news event like the cloning of Dolly can cause people to reflect on local effects.

Gamson's focus group used *conversational resource strategies*. There were three of these: *media discourse*, references to events, known catchphrases, or public figures from the media; personal *experiential knowledge* from the speakers' past; and *popular wisdom*, which were wider, oft-repeated cultural phrases of undefined source that may also be expressed as catchphrases or maxims. Popular wisdom is a particularly difficult category to define. It can be part of the expression of a subculture: 'We've always been discriminated against'; 'You can get killed crossing the street' (ibid. 124); or catchphrases that enter into popular culture and conversations, such as the prevailing one in this study, 'playing God'. Popular wisdom also has, according to Gamson, cultural resonance that 'condense and concretise' the themes (ibid. 143). Gamson was also involved in a comprehensive study with Ferree *et al.* (2002) of how abortion discourse was framed differently in the US and Germany. Content analysis of newspapers showed four different frames advocating rights for reproductive issues, either for or against the embryo.

Care needs to be taken in directly comparing young people's RGT frames with news media. Although the field research reported here show signs of their being media-savvy, empirical evidence from their pre-discussion responses and teacher quotes suggest that young people are not always *news* media-savvy. However, in school conversations it is not so easy to disentangle the experiential from the mediated, the individual from the collective, the internal from the cultural, or the private from the public. Gamson (op.cit.) tells the story of a friend who discovers

Table 6.1 Recurring phrases throughout the study that suggested a popular wisdom (Gamson, 1992)

Nature
'Messing with nature'
'Playing God'
'Let nature take its course'
'Everyone would be the same'
'It's not natural'
'That's just wrong'
'You are what you are'
Science
'Standing on their shoulders'
'Might as well [use PGD]'
Life-planning
'You might as well have a full life'
'Give him the best life they can'
'It's not fair on the child'
'Turning childbirth into shopping'
'What's meant to be is meant to be'
'Looking for perfection'

to his dismay in adulthood that a treasured childhood memory was in fact a scene from a film.

In Chapter 2, we placed into a brief historical context the biotechnological applications emerging from the late twentieth century and moving into the twenty-first. The announcement by Crick and Watson in 1951 of the discovery of the DNA double helix is considered the beginning of molecular biology and genetics as a major technological discipline. There are many ways today that people might come into contact with these biopolitical streams about genetic information, technologies, or actions that are perceived in some way to change or control the human body. They are likely to come from strategic outreach programmes, or news bulletins informed by such programmes. Publics can access 'elite' biotechnology discourse either through scientific papers or websites (National Centre for Biotechnology Information, 2013), and open access is an ideal not yet realised, much to the frustration of advocacy groups, the real life political parties'.

How comparable are debates in the classroom to those outside? Pro-life or pro-choice extremists are well aware of their rhetorical, as well as their symbolic, power. The scientist has the symbolic power of authority. The NGO has the symbolic power of friendly and healthy nature. Part of the ethnographic approach involved looking for clues to outside influences that show young people have some awareness, if not well informed, of both the science and the issues. This awareness of 'genes in the ether' feeds into identification with a young person's struggle with genetic disease, an embryo's humanness destroyed, a character's plight in a film.

The films reported throughout this book created one of the primary spaces for debate. It is informative at this point to briefly look at another mediated example to show the performance dynamic of televised experts. In 2005, following the publication of the CAHR report (Commission for Assisted Human Reproduction, 2005) mapping out how legislation on the embryo might look, the RTE flagship current affairs show *Prime Time* (2005) had three high-profile guests debating RGTs: Maureen Junker-Kenny, Associate Professor of Theology at TCD, arguing against, Prof. David McConnell, the geneticist arguing for, and Carol Coulter, Legal Affairs Correspondent with the *Irish Times*, the neutral legal representative. Professor Junker-Kenny's frames were medico-legal (McDonnell and Allison, 2006), but she also attempted to present the argument through the eyes of a future child, should PGD be legalised, and endowed such children with a sense of life politics: 'If I am not good enough for my parents then they're not good enough for me' or 'certain lives not worth living'. She aligned her position with a Commodification frame once Professor McConnell also began using a Future Child frame.

It could be imagined that Junker-Kenny began to cede moral victory in the eyes of the viewer as she simplified her 'dignity' argument to the 'reduction of person to body and from body to genes'. Her visual technique may have been conceptually effective but may have been just that, too conceptual, too academic while McConnell went for a more culturally resonant set of principles, speaking of the 'tragedy of infertility eased by IVF' and the 'humane' choice of easing a future child's suffering. McConnell, as well as using science as the moral authority – here Coulter's legal advice was the objective voice – embodied the impatient, deficit-model scientist, almost arrogantly dismissing Junker-Kenny's suggestions as inferior. However, by saying 'very few people get into this complicated discussion', he may have seemed a little patronising, but he was correct. At least, in the common-sense world that young people in discussions were inhabiting, whether they agreed with McConnell or not on his RGT stance, they were not as willing to use the terminology of 'the nature of parenthood' or 'future vision of how we want to live' that were the core of Professor Junker-Kenny's discourse.

The dynamic from the classrooms was all here. McConnell was the protagonist, as were school participants such as Derek (NDC1) and Kate (SK3). The TV panel utilised arguments from scientistic as well as personal frames and using body language of the moral victor. It was an aspect that revealed a curiously Irish response in young people's almost detached mantra of 'you might as well lead a full life' or 'you might was well prevent it', perhaps allowing the very existence, the very power, of the technology itself to be its own justification for using it.

A living Likert: choosing positions of agreement on film characters' decisions

When students were asked in the field studies to choose their position in relation to decisions made by film characters, their responses can be seen as moving into the discursive, presenting their opinions on the use of PGD, their views on science, religion, identity, or society. These performative, role-playing activities have a particular context though. Participants were responding to film characters and to pedagogic power plays in the classroom. The analysis in this chapter contributed to their motivations for choosing positions on the statements, will tease out more comprehensively their frame constructions about RGTs and how they relate to a background of discourse in Gamson's (1992) terms. The positions taken are a form of role-play. The local, performative strand taken in this chapter derives from Goffman's more dramaturgical work (ibid. 1959) (which can also be seen in the positioning analysis of Harre, Davies, and van Langenhave (Davies and Harre, 1997; van Langenhave and Harre, 1999) and Bakhtin's explorations of spatial subjectivity and physical positioning (Kazan, 2005 citing Bakhin, 1981).

Positions are more dynamic, according to Harre and van Langenhove (1999) paying close attention to the local moral order, the local system of rights, duties and obligations, within which both public and private intentional acts are done (ibid. 1).

Physically, the students in the 2005–2007 study stood in position along an invisible five-point Likert scale of agreement in relation to others, a 'living Likert', or a 'moving debate' as a youth services facilitator in the pilot phase called it from previous encounters he had with this method. It is important to note once more, as discussed in Chapter 3, that this Likert scale was not meant in the final analysis to represent absolute views on the particular technological application that was the subject of discussion. As has been addressed throughout this chapter, the dynamics of performing opinions themselves shed some light on social interaction and peer grouping around issues.

While considering the context, it is still instructive to take a closer look at the common agreements and disagreements across the schools on issues that are part of common discourse. The most common agreement on statements were those relating to: the rights of young people to be tested (without an arbitrary age limit, but on the basis of maturity and whether or not someone is already suffering from a genetic disease in the family); preventing a child from getting a disease even if that means stopping an embryo from being implanted; disclosure of sensitive information to patents known to scientists and clinicians; being uncomfortable with sex selection and selection for traits other than those that are life-threatening in PGD; the need for boundaries for bioscientists' practices (moral and legal); the need for boundaries for everyone regarding what is or is not acceptable in RGTs; the belief that people's best intentions are generally at work while making decisions about RGTs, but that does not make it morally right.

There are many contradictions among these agreements, reflecting the complexity of the moral arguments facing people having to make decisions about these technologies. Personal motivations were presented in both dramas. Particular disagreement also seemed to exist about regulation, in terms of where responsibility lies, and much of this is evident in the rights frames discussed in the next chapter. There was also ambiguity regarding the idea that the protagonists were 'doing their best' for the 'beneficiary' of the technology. While many saw the best intentions in characters' actions there was still uncertainty among others whether Jennifer and Ryan were doing their best for Mark in *The Gift* when they selected an embryo by PGD or whether Alex was doing her best for Andrew in *If...* In arguments, the most degree of certainty came from, perhaps not surprisingly, those who rejected the use of the embryo in any technologies. However, these arguments were largely in the minority.

What comes out strongly however, and the theoretical reasons for this were analysed in Chapter 4 – was that peer power was a major factor in how people responded to film characters' justifications (although in all schools for Statements 2 and 4 of *The Gift* (see Appendix) there was generally unanimous agreement without peer power). In other words, there are great structural reasons such as belief and codes for young people to step into discourse. But, the immanent power of relationships – friends, boy/girlfriends, fancying a classmate – often won out when a student made a decision on 'what side she was on'.

The troll

Regarding where young people take resources for moral decisions, even those that are from the mundane and the everyday are informed by cultural norms that are increasingly media-influenced (Boltanski, 1999; Couldry, 2003). How the morality (or conversely, the ordinariness) of RGTs are discussed in class will depend very much on the medium of discourse, in this case, film. There are code of practices from courtroom dramas (*If...Cloning Could Cure Us*) and studio debates in film discussions that influence the way the discourse is shaped, for example a media audience or Q&A in the bioscientist session.

The word 'shill' as used by Goffman (1990)[1959] referred to a performer who manipulates audiences and other performers into action, but contemporaneously is secretly associated with a strategic force. Today, on internet forums, where there are anonymous contributors to loud debates, there are lurkers and cyberbullies. There are also trolls. A troll may be a shill, but her vested interest is of no importance – she is there to turn up the intensity, cause mayhem among those too close to the centre, who now like the pigeons startled by the cat, scatter opinion to the extremes. Derek from NDC in my school research was one such performer, arguing for the sake of it and creating straw men of opposing arguments (there were few others across the six schools, except the odd Nazi sympathiser, playing with controversy). However his high status was taken from him when he seemed to make a gaffe by implying that abortion referred to a foetus close to birth only and not the destruction of an embryo. Thereafter, more participants turned against him in arguments as he had lost his authority. Karen from SK also provoked strong reactions for her pro-life views. Very often she stood alone in protecting the embryo's voice. I felt the need to help with her defence and to control animosity rising against her, as for her, it seemed to go beyond a 'game'. Karen may have been someone for whom the debate infringed on personal beliefs and so, as teachers have done before in socioscientific argument, I tried to protect her from a life political issue. Des also from NDC, by contrast, was also pro-life, but was quieter in debate. As Karen was provocative, Des was sheepish when putting forward his views. Karen drew criticism, while Des had sympathisers.

Steve Fuller, by agreeing to participate as an expert witness against Tammy Kitzmiller and in support of Intelligent Design as explained in the last chapter, was a highly rhetorical and intellectual form of trolling, for the sake of thinking the biosciences anew.

Summary

Is framing just another word for confirmation bias? I have shown that it is a controversial methodological area, but nevertheless it is carefully considered in this section, for framing strategies, tactics and conversational resources – and their practical use in classrooms. Young people, in common with adults, 'frame' and 'position' themselves in a discursive universe about highly technical, technoscientific matters of concern. Whether moving on the classroom floor or through social media, once they frame and position for or against RGTs, they have already entered politics.

Notes

1 Technology was also proximal for Heidegger 'nearness' and this is part of journalistic code.
2 Gamson is not within the 'practice theory family' yet his media communication and cultural frames are methodologically useful here to connect the practices of the local and the global without recourse to cognitive individualism or intersubjective transcendentalism.

Chapter 7

Nine genetic futures

We have now engaged with compositions that organise structure and dichotomies that sequester life politics. *It will take much shouting to proposition life political argument.* This chapter opens the door and begins the shouting. I present nine potential futures for genetic engineering/screening of humans, by way of discursive engagement between students' own accounts and the philosophical, rights-based arguments of scholars, particularly in both science and media studies, on how future subjectivities may be affected by RGTs. The intention is to show that there cannot be a solid Leviathan (Hobbes, 1651) or prescriptive social contract (Rousseau, 1998 [1762]) as we deal with ghosts or phantoms of public opinion (Lippman, 1993 [1927]), or governance and law (although we see how Rawls might give us a way forward in the final chapter), but entering into debate has created these political entities worthy of representation. We saw in Chapter 3 how and from which sources public (and student) 'opinion' is formed, and in subsequent chapters we explored how misleading option and beliefs are when investigating this.

What would representations to a *futures* parliament look like? If each child who speaks had control over RGTs, would these voices from the parliament emerge? The futures represented here have been given the name 'frames' and 'themes' in my research in the past (Murphy, 2005, 2008), but I have dealt extensively in the previous chapter about the problems with descriptions of media/performative frames. Rather the future framings are ideas of the future that privilege an entity or set of entities based on classroom debate, but bolstered by both STS and bioethical readings. Science education research, continuing the argument from the previous chapter, can learn much from this. Using a type of frame analysis – constant going back and forth over transcripts from schools, arguments of frames around identity, mobilizing, mini-social movements – the futures created here are co-produced by the media research literature, but also film, mediated cultures of *bíos* and most influential of all in the analysis, young peoples' visions of near-future humanity. The thematic configurations were based on how participants organised others around a semantic and cognitive core. They were created by line-by-line searches for metaphors, repetition of words, shifts in content from audio transcripts of the events. They are the organisation of words and concepts that people will do in a situation from linguistic and sociological traditions (Ryan and Bernard, 2003), or 'personal semantic networks' (ibid. 281). 'People [frequently] circle through the same network of ideas' (ibid. 281).

The nine genetic future framings are: Modern Progress, Harmony with Bifurcated Nature, Patient/Parent Autonomy, Sacred Embryo, Future Child, Future Humanity, Fortuna/Fate, Commodification and Pandora's Slippery Runaway (each a development of Gamson and Modigliani (1989)), but also Hornig Priest and Ten Eyck (2003).

This chapter is the real operational link between the school empirical work and future life politics. It attempts to (re)establish that link, fill the gap already identified in the book, between the cleaned-up curriculum and messy nature of everyday life (processes), and its future politics. It brings into focus futurescapes, prospects rather than defined futures (Latour, 2010), forging futures with agency of one particular person or group. They are prospective – they certainly are not absolute.

The futures created from the accounts that follow are pre-dialogic *ad hominem* arguments contributing to what Latour calls the 'speech impedimenta' (Latour, 2004). In the end, the arguments lack bio-content knowledge at times, but how much is needed? This will be part of the analysis.

'Modern Progress'

The Modern Progress framing mechanism identified in young people's conversations connects to the cultural theme Gamson (1992) identified as '*progress*'. The term I use is also an amalgam from Latour's use of the 'Moderns', those who insist we remain on a linear trajectory of science, technology and innovation from the Enlightenment. There are, as have already been demonstrated, occasions where participants in our focus group session show that young people have learned no other way in science. It is their expectations of a scientific subject, generally unpolluted by (or immunised against, to use Esposito's biological analogy) videos and discussion.

A participant who uses the scientific rationalism frame for an entity – human or non-human, collective or individual, factual or fictional – may refer to a need for technology to remove a disease; or describe the process and connecting issues in terms of scientific practices only; or refer to the entity in de-humanised language, the exemplar here being 'embryo'. It is a form of Horkheimer's 'instrumental reason' (Giddens, 2012 [1990]: 116), that is impossible without an institutional sequestration of experience (ibid. 1991), an institutional reflexivity manifested in a young person's deference. In other words, the speaker de-cleanses of morality and aesthetics. It is a normalising frame in line with the dominant paradigm of scientific discourse (Hilgartner, 1990). All progress ideas are for betterment, a linear trend – some students who knew their evolutionary theories used this trope for a narrative of betterment – but ultimately science (capital S, in its Nature-revealing guise) is the driver. It is strongly linked to scientism. Scientific schemas were common, which is not surprising given the activities generally took place as part of a science class. I have argued elsewhere (Murphy, 2005) that Driver *et al.*'s three epistemological representations, science as content, process and social enterprise are fundamentally integrated, but content and process are the strongest, maintaining the borders of Modern Science. Ryan and Alex, the scientists in *The Gift* and *If...Cloning Could Cure Us* respectively, were the exemplars of Modern Progress, yet they both pushed the envelope – each stepped outside the ethical practices and self-governance of science for 'more human' ends (eradication of FRDA for Ryan, development of spinal therapy for Alex).

If we look at young people's responses, Modern Progress suddenly becomes impossible without strong links from other regimes of justification. Damien from NDC provided a sophisticated version of science, remarking that there was a 'God replacement' in action, not entirely different to Midgley's (1992) book *Science as Salvation*. 'It's like....fixing people's lives,' one girl said of RGTs in ND. 'You know – you can have normal babies.' Modern Progress has, then, a moral, social and political role, not too dissimilar from the old social order arguments of Hobbes, or indeed Popper' 'piecemeal social engineering', a Christian, Calvinist, technology leading to the divine (Fuller 2011: 83). One student saw biotechnology as 'the moral side of biology' (ND, female, interview).

When participants in the sessions with DCU bioscientist speakers were invited to ask questions, their first questions tended to be: 'when will they find a cure?' (Both speakers dealt in some way with cancer research.) Most participants across all schools saw PGD as preventative medical intervention above everything else, even where there was a moral issue in the use of technology. In defending its use, many referred to how biotechnology would change lifestyles and how PGD in particular could provide a 'full life' by selecting for 'important diseases'. This idea of a 'full life' I will come back to again throughout the chapter, being as it was used extensively, demonstrating the prevalent use of the term as a possible example of Gamson's (1992) popular wisdom, what 'everybody knows', a device to garner *argumentum ad populum*.

Fundamental discursive concepts such as 'science' and 'nature' were placed, in the analytical approach taken here, into the context of framing power and moral practices. Douglas from SD – who often demonstrated an acute awareness of the nature of science and technology – saw biotechnology as 'controlling' biology. In certain schools, notably participants from ND and SW, the rationale for PGD was a black and white issue. 'Have the disease, or not have the disease. End of story' (ND, female, *The Gift* film discussions). As noted elsewhere, I observed in SW responses a certain lack of impatience with the line of questioning on impact to society or identity that added to the complexity of the issue. The 'common sense' applied here was disease prevention. There was often a clear moral distinction between technology with the power of prevention and technology as controlling nature, human identity, or human affairs. In 'selecting a child', using PGD did not have the goal of 'mapping out its life', as one participant from the all-male

NDC discussions said. It was preventing him from having a 'horrible life'. In all schools however, students advocating the technology without moral constraint were in the minority.

In classroom sessions, there was some genetic determinism and reductionism. Why would it be otherwise given media focus on 'the gene for' (Chapter 2) and the fact that our conversations focused on one gene that was presented as being the cause of FRDA? Screening an embryo meant screening for a gene. There was a certain amount of enculturation by the film here. Annie changes her lifestyle to accommodate the physical effects of ataxia because of 'this wretched gene' (Baldwin, 1999). In natural conversation, we cannot help using a single gene to represent a condition or trait. It appears to be another type of cultural resonance. This was demonstrated as we continuously referred to 'the gene for sporting ability' or 'the gene for Friedreich's ataxia', perhaps prompted by the simplistic nature by which the film portrayed genetic predisposition. It was natural that participants did not generally have enough information to argue if genes have that much influence on disease. They were never aware of any wider nature/nurture or systems biology debate.

Yet nature/nurture ruminations occurred and the topic was touched on in *The Gift*. Mark, the teenager who realised he had been screened, is upset at the end of the film when he realises that an attempt was made to pre-select his destiny by selecting an embryo. This may well have presented a new phenomenon to most students – a distinction made between genetic make-up and a person's own feeling of identity. Although many professed to understanding why identity would be threatened by PGD, others did not always see Mark's problem. This was the general feeling, until a distinction was made between Nature and identity, a realisation of an individual intrinsic essence present outside the genes. For boys in our discussion, particularly, this still appeared to be ingratitude – why would Mark complain, if he was given this 'gift'? This observation demonstrates a certain consequentialism (Hayry, 1998) and correlates with males' likelihood of accepting genetic technologies as described by Hoban *et al.* (Whitelegg, 1998 citing Hoban *et al.*, 1992).

The film – and therefore our co-constructed reality of nature, disease, and intervention in film discussions afterwards – also focused on the either/or scenario of leaving nature to take its course or take some action to prevent a disease. In so doing, as we will explore in another way in our account of the Fortuna/Fate framing below, this disease was seen as almost inevitable once an embryo was implanted which contained two recessives genes for FRDA (to once more use a monogenetic reductionism). There was, however, a confused relationship between this genetic essentialism and the identity that was to become 'Mark' even in the film itself. Ryan tells his PGD-selected son Mark: 'You probably would have been in a wheelchair by now if your mother had anything to do with it' (Baldwin, 1999). This was a common 'misinformed' idea of how the PGD technology works. In the pilot, I identified three different concepts of embryo selection as represented by participants (Murphy, 2005):

- as a means of preventing disease in the same, pre-destined person. Mark would have been born anyway, just with different traits;
- equivalent to the removal of genes from the same, pre-destined person;
- as a means of preventing disease leading to the birth of any one of a number of individual people.

The last interpretation is the scientifically accepted version. Fundamentally, this is how PGD in the arrow of time works. It was curious how the first two understandings above emerged among these participants. To at least three participants who voiced an opinion from ND and NDC, Mark would have been unhappy if he had the disease and then found out that the technology was available but had not been used to 'remove' FRDA. It may be a natural instinct for some to apply the normal rules of pregnancy to PGD even when there is a basic understanding of the technological process. The old reproductive modes of thought may be difficult to modify. At one point, Denise (ND3) doesn't get this, and is corrected by her classmates.

ND3: Say if you're born right? And you've a disease right…and your ma turned around to you and said I could have stoppin' that but I didn't like…
M: You wouldn't do it.
ND3: You obviously gonna think yeah then you should a just stoppin' it.
S: Yeah but you couldn't have known!
ND3: …could have had a normal life if I had of told you would obviously gonna think oh nice one!

(ND, *The Gift* film sessions)

Denise was eventually told 'it would've been a different kid' (ND, female, film session). In the language of clinical science, as with the fertility doctor utilised in *The Gift* when supplying Jennifer with her genetic test results, and indeed the conversations I had with Beacon CARE Fertility, prevention is a matter of choice for the future parents and where the Modern Progress framing is invoked, very often this is augmented by a social responsibility theme (see later in this chapter) or welfare argument focusing on the future child. The parents have the *facts* in their possession, now it is time to *act*. 'Do something', as the Wellcome Trust statement has it (Townsend, 1999). In the diagnosis scene, both Ryan and the doctor are involved in a gentle act of persuasion with Jennifer to get her to see the logic of PGD. However, the embryo that eventually became Mark was selected as a result of the action of this one person; Ryan, a geneticist who could 'read genes' acts alone, and clandestinely, against Jennifer's wishes. In the words of one ND respondent, he may well be 'into genetics so he knows what the story is' (ND, female, film discussion), yet in a deceitful way, his knowledge becomes part of the expert systems that become Jennifer's aggressor. Some SK participants, defending Ryan's actions up to a point – they agreed he should have chosen a disease-free embryo but not on the basis of other traits – opined that human nature being what it is, Ryan would want to do what he could for his future child and thus go further in selecting the 'best traits'. Therefore in this case scientific knowledge militates against the right thing to do. Scientific knowledge carries a heavy responsibility. There were often participant statements like 'I suppose it's wrong but I'd do it anyway'. This notion breaks from a normative moral decision.

For those who argued that Ryan was only doing what was 'for the best' – and the amount of these respondents was considerable – scientific knowledge is the defence offered for the decision and action. Even in acknowledging the performative nature of the following exchange, it is interesting to see how far the Modern Progress frame can be carried. This is the male NDC group interacting with Derek, the laid-back rebel from that class:

PM: OK, so for yourself there, would you have done what … Ryan did?
L: Yeah if I had the chance.
NDC1: Do what?
B: I wouldn't.
PM: Select a boy… OK select a girl sometimes…but select somebody to their best, what you think would be the best in the genes.
S: Yeah/Yes.
PM: And if you could read genes… OK what would be the person you'd pick?
NDC1: Ahhhh, it'd be someone not conventional sport but good at…
PM: Good at what?
NDC1: Good at academics like… intelligent like.
PM: So an intelligent person who's good at sport.

(NDC, *The Gift* film and discussions)

There is a performed genetic determinism active here, as the boys get carried away, many accepting the 'best genes' argument when presented as a question on the kind of person they would like to 'create'. This is a contradictory position for those who otherwise would not set themselves other deterministic opinions, such as Derek, above, as previous quotes show. It is part of the performance, part of how pedagogy carries along, *dispositif* and *habitus.*

The objectifying of embryos and genes as sites of disease brings an instrumental effect to identity – remove the gene means remove the embryo means remove the person. The application of Modern Progress, particularly when incorporating elements of risk rather than fate, can reduce human identity discourse to genes. The lottery of birth, whether explained in 'natural' terms or by a new technology, denies the existence of a 'future child' that was not likely to be born in the first place. These explanations are becoming more prevalent in the role of family and relationships generally in modernity. Franklin (2003), in her anthropological work, sees that the ties of kinship and responses to natural facts-social facts distinction are being changed by RGTs. The philosophical implications are not lost on Damien:

> I think when you're going through life…you're going to come across these philosophical issues yourself. And I think any help you can get in that whether it be through biology or religion class or whatever to help you cope with that is well needed. And also facts in a subject can only get you so far. But until you start putting something into the class and you can see how it's affecting like life…everyday life….really then the subject starts to …come across strong in it.
>
> (Damien, NDC, interview)

This messiness of application and process results in what he calls a 'grey area' of scientific endeavour outside the 'black and white' distinctions of science content. 'Grey area' was also a term used by Des from NDC in a specific bioscience application: the treatment of disease. It is here that science becomes associated with human embodiment. It touches the sacred and there will inevitably be clashes of views, according to Des.

The fact that controversy exists within a discipline like the biosciences at all was raised by Kate from SK. Kate was considered a high achiever in all subjects, but opposed the form and structure of her education. In an example of the conversational resources that students had at their disposal, she contrasted the conservatism in biology education, with her other passion, art.

> You couldn't talk a hundred years ago about the nakedness of art and stuff like that. That was controversial. But there's nothing much anymore.
>
> (Kate, SK, interview)

Kate's rationality is absent from the frames created by many others however. She explores her worldview on the question of religion:

> I don't know… I don't know what's out there but I don't believe…well… I'd like to think that there was something else out there rather than just the Earth, being that… It's not logical I don't think
>
> (Kate, SK, interview)

In *If…Cloning Could Cure Us*, the central controversy revolves around the use of a 19-day-old embryo for stem cell research to treat a spinal injury, exceeding the 14-day limit permitted by British law. Participants from LH challenge the futility of an arbitrary date of 14 days based on science:

> *Y:* I think like I don't...I disagree with the 14 day. I think they broaden it out to maybe 19 days…Essentially it's the same as a 14-day-old embryo. The only difference is differentiation.
>
> (Female, LH, *If…* film discussions)

Using the process of science to decide when human life begins or differences between the embryo as human and the embryo as artefact is currently a live one in Ireland. Jasanoff (2005) has criticised the Warnock Report in the UK for instrumentalisation of the embryo by fixing the 14-day threshold, as though this temporal distinction could decide between human and object (Jasanoff, 2005 citing Warnock, 1985). 'Y' above shows herself to have at least been informed by the film's representations of science regarding differentiated cells and draws her own conclusions about how useful this is in dealing with the issue. However, in the next sentence, Y contradicts herself by saying perhaps 19-day embryos would be more suitable after all.

'Harmony with bifurcated nature'

Those participants who use the Harmony with Bifurcated Nature future framing may not necessarily reject a scientistic rationale but will reject any attempt by a pro-technology framing to downplay the negative impact of technology on their concept of nature or natural practices. Again, this category is modified from Gamson and Modigliani (1989), a play on the 'harmony with nature' frame used quite frequently in media frame analysis literature, but also invoking Latour's constant references to Whitehead's bifurcation of nature, the easy dichotomisation of Nature and Culture. An example from *The Gift* Statement 1 (Townsend, 1999) at the North Dublin school (ND):

PM: Yeah, but if you thought that about everything...
L: Yeah but if all that stuff is going to happen you might as well just like, stop it.
O: Prevent it.
P: That's messing with nature though.
L: You can prevent it from happening...
N: That is messing with nature.
L: Obviously it is.

(ND, *The Gift* film discussions)

The phrase 'messing with nature' is used constantly in this ND discussion. The prominence of the term 'messing' in responses suggested the naming of this frame. Once students latched on to this frame, other considerations – either the rights of the future child to a healthy life, or easing suffering, or the case for the extraordinary life-planning measures of Ryan for his child – were often forsaken for a common impulse against the idea of 'tampering', either in the destruction of a living entity for research or the idea that a parent could be 'picking all your embryos and all'. On other occasions, other frames were aligned with it to support it. At times young people used secular arguments against the use of RGT, or other biotechnologies, that extended beyond any sacred understandings of the embryo. In discussions, the concept of nature was introduced by the words of Statement 1, yet it has a powerful salience by drawing on a disruption of normal practices, appealing to a higher law that does not require theistic explanations.

Participants who responded that Jennifer 'wanted a natural child' meant a child not selected by PGD, thus bestowing implicitly a kind of artificiality upon a child that would have been born using this procedure.

Of the interviewed participants from schools, Des from NDC, Karen from SK, and Lorraine from LH had the strongest negative reactions to using the embryo in either research or for screening purposes. (I cannot place too much emphasis on the fact that NDC and LH were Catholic schools, although it must be noted.) Other reactions were evident against using stem cell technology or, more commonly, using PGD to select characteristics such as sex, physical prowess, and hair colour.

Distinct lines are drawn, which correlates with Jasanoff's (2005) suggestion that controversies at the interfaces are highlighted to prevent boundaries becoming blurred by the structural intersection of disparate normative processes. Across the responses to the eight statements of the Wellcome Trust debate activity (Baldwin, 1999) based on *The Gift*, it was evident most young people in the six schools did not think sex selection should be permitted, nor should selecting particular physical attributes if not related to serious disease. However, for many others, the mere fact that an element of autonomous selection is available at all draws a kind of nervousness. For others again, the use of the embryo goes beyond concern to a type of revulsion, the yuck factor again. It is tacit, instinctual, perhaps at times beyond words, but the yuck factor nonetheless may explain why certain practices in any given field are almost self-guided within their normative pathway. The arguments against tampering were often settled at 'it's just wrong' or 'no, that's not right'. Although one ND student (jokingly?) made a reference to babies in test tubes to create an image of the non-natural, the females from this school made a distinction which suggested being unnatural was not in itself the issue. 'Sometimes going against nature isn't bad,' said one. 'Like taking a tablet. But when it comes to genes...' (ND female, *The Gift* film discussion). Participants reacted strongly when one person suggested taking an antibiotic could be analogous to using PGD if the nature argument were to be used. The more analytically minded who framed RGTs as meddling searched for logical framing. 'Balance' was Damien's example of an attribute of nature that was normative and normalising and one of those 'species of animal' in the kingdom of nature – 'man' – can 'upset this balance' through technological intervention. Lisa from LH presented another hypothetical scenario that had ramifications for the natural: if stem cell research was used to save someone from dying, as in the case of Andrew's in *If...*, then the extension of this would be the prolonging of a person's natural end. Could this be seen as contravening natural law, she wondered?

Damien's unusual combinations of the two frames, scientific rationalism and against messing with nature, was further called upon to argue that intervention as prevention was the lesser evil, but warned that 'there's prevention and then there's... getting too much involved like' (Damien, NDC, interview).

Damien does not favour PGD, however. He uses a scientific rationalism frame to explain this: the 'gene pool' may be 'upset' if we humans tamper with our own processes of natural selection. In external debates, it is common to have scientific arguments against RGTs. We can see a value system working here in the language of science.

Some scientific rationalism was represented in a tactical, rebellious way, as opposed to the dominant strategy (de Certeau, 1984). The subsequent court case in *If…* transpires as Alex's colleague informs the authorities of her procedure of using the forbidden embryo. Yet she has some support from the audience here, some of whom feel that normative practices of science were not enough in this case and whistle-blowing – or 'grassing' – prevented a cure. Science needs rebels in this consensus. It seems peculiar that a consequentialist risk frame also emerges here in the argument: whatever risk there may be from selecting an older embryo eventually 'balances out' in the moral scheme if a cure is to be found. Is this a scientific risk or a moral risk? What was further questioned about Alex's act was the immanent nature of her own desires: did she really do it to help (or save) Andrew or did she do it to further her own career? Lisa from LH, reflecting on this discussion, tells the group that Alex's mixed motivations are understandable – fame and fortune are appropriate goals if the good intentions of life-saving therapy is also the outcome.

By and far the most utilised popular wisdom term that might exemplify Harmony frame's appeal to higher moral law is the phrase 'playing God'. The perspective is well covered in literature (Goodfield, 1977; Kass, 1985; European Commission, 1997; Bauer and Gaskell, 2002a; Verhey, 2003). This metaphor alone would be the subject of a book such is its cultural resonance. While once or twice I introduced this term myself to watch it 'take off' there were other occasions where it emerged spontaneously from student discourse.

'Fortuna/fate'

Themes or schemas which drew on ideas of Science and Nature to explain RGTs related discursively to *risk* and *fate*, two oppositional but oddly connected framings, strongly tied to how Giddens (1991) engages late modern self-identity in the context of technological change. In the dichotomy of Buchanan *et al.* (2000) fate leaves things to chance, risk offers choice. This draws heavily on Beck's (1992) risk society, but also Giddens' (ibid.) Fortuna/Fate versus risk, the tradition of Greek foreboding of fate, the structural futurescaping of political action. Fuller (2011) maintains it is Divine Providence as passed down from Protestant history. Roman Catholic teaching, as these Irish schools represent, might tie less to Providence, but fate is still there; in fact it was part of Thomas Aquinas's teaching. Rose (2007), like Giddens, sees that control, deliberation and careful life-planning decisions now break from this fate. Indeed genetic susceptibility is a Modern version of predisposition, a pathology or darkness within ready to pounce (ibid. 19). It links heavily with degeneracy, requiring immunisation (ibid. 19, also op.cit., Esposito, 2008). (Aquinas knew we had some free will to fight the *acedia*, the inner structural foreboding (Fuller, 2011: 97).) *Fortuna*, according to

Macchiavelli, is the 'arbiter of one half of our actions' (Macchiavelli, 2005 [1532]: 84), although this was a rhetorical statement rather than statement of fact. But this half has shrunk immeasurably, goes the narrative, by the knowing and controllability now at our disposal once predispositions have been assigned; we can counter-plan, continuously assess, self-improve, buy our way out of what Fortuna has planned for us and break from our predestination. It is also moral conduct for future generations, Foucault's biopolitics writ large. The politics of choice. Now that you know, what are you going to do about it? The 'gene for' phenomenon also allows better prediction, better control. Rose (2007) demonstrates how this susceptibility (information) turns easily to enhancement (action). In the movie I presented in classrooms, *The Gift*, the risk is assessed, the percentages given for being carrier/potential victim supports risk assessment. When an expert system – in this case a PGD clinic – presents this risk to Jennifer, according to this argument it is the power of facts that is persuasive, with doctor as bio-orator, representing the structure of fate sent from the lab. There is a *common sense* allowing the potential parent to colonise her future. It should also be noted that those who argued their case here often focused on the core premise of statement itself, the *right to argue* in this Parliament of *Fortuna* and Risk ('Was Jennifer right to argue to let nature take its course?' (Baldwin, 1999)). If Modern Progress favours risk assessment though, Fortuna remains:

F: ....like if they're meant to be born that way they'll be born that way and she'll still love the child.
M1: Yeah, but then they'll only live to a couple years. You might as well have a full life.
F: You can live to forty, fifty, you can.
M1: But that's just not going to happen.
I: But you can be downed by the back of a bus at two.

(ND, *The Gift* film discussions)

The phrase often used here, particularly by ND participants, is 'meant to be'.

> I think ...just let nature take it's course 'cos some people are meant to be good at sports, some people are meant to be good at academics.
>
> (Male, ND, *The Gift* film discussions)

These arguments about a child's destiny are particularly pertinent when confronting parents' future life-planning for their child, as will be clearer in the rights framings and Commodification sections below. Where there is a right, parents are perceived to break with normative practices by controlling fate. Derek, in his earlier nature/nurture ruminations, seemed to invoke a risk frame in contrast to Fortuna/Fate when talking about control of the next generation, also bringing in crude manipulation descriptions of masculine identity and personality change:

NDC1: The whole thing right. If you pick someone with attributes, they can still fail at life… It's still an accident that you're good at sports. And someone with a high temperament you can still fix it.

(Derek, NDC1, *The Gift* film discussions)

Yet while Derek was alone in a later interview, he decided that Ryan should have selected an embryo randomly once it was clear that it did not contain any genetic markers for FRDA, thus introducing Fortuna back into the process. While the inevitability of disease when present 'in the genes' suggests genetic determinism as mentioned earlier, the element of surprise is crucial to appreciating a child for many who frame childbirth in terms of chance, but with elements of Fortuna. The action of PGD acknowledges risk but resists Fortuna.

K: And there's no surprise, like. You know what way they're going to be born. You're not going to see which way they grow. He already knows which way... He's not going to get any surprises like.
J: That's the best thing about having like ...a child grow up. Is seeing how they develop. And what way they take to things.
K: Yeah and he's already going to know.
F: And deciding what they're going to do in the future.

****

L: Yeah but if all that stuff is going to happen you might was well just like, stop it.
O: Prevent it.

(ND, *The Gift* film discussions)

Two girls in ND were of similar mind about risk:

ND3: (Denise) It'd probably go wrong.
C: Yeah, there is a risk that it could go wrong. Cos everything carries risks like so…
You could end up picking your child not to have Friedreich's ataxia but there could be something they don't know about, that it would have, that you don't know about either.

(Females, ND, *The Gift* film discussions)

The counterargument goes then that, rather than Fortuna, risk and the uncertainty of the future alone were enough to allow 'nature take its course'. As one girl said, it would be pointless to get rid of the disease since something else could happen.

Living in a risk culture does not mean previous eras have enjoyed less risk of harm or discomfort to the body on a day-to-day level. However what is unprecedented about late modernity is the integration of statistical risk into life-planning (Giddens, 1991; Beck, 1992). Institutions create the parameters and manage risk (ibid.). The ND participant cited above may be defeatist and pessimistic in her attitude to PGD on the face of it, but in the context of her defence against 'letting nature take its course' in discussions, this framing shows awareness of the risk of the unknown.

The concept of time, and particularly the future, has preoccupied sociologist Barbara Adam (1990); how we humans perceive ourselves in the future, as old with children, and for young people, with careers and adventures. Groves calls future-oriented responsibility a type of non-reciprocal responsibility, a type of care for technological futures (Adam and Groves, 2007). It would be a dialogue rather than a colonising of the future. A Modern Progress future would itself have something similar, Natural Law predicated on reciprocal altruism (Fuller, 2011: 89). What the risk is a continuation of Galton's project – we need to help Nature along towards God, via a eugenics regulation to the point of coercion (Fuller, 2011: 194), what Rose (2007: 55) calls the 'science of improving stock'.

'Parent/patient autonomy'

The future framings over the next few pages are rights-based. If we contrast liberal individualism with social responsibility, we see loci opening up here in young people's discursive actions. Young people were only too aware of the locus of control, how these technologies becomes a control of our sense of self and destiny. The possible emergence of a liberal individualism is another theme considered by Gamson (1992). The tendency towards liberal individualism is characterised as a rational subject shaped by personal will. Although it is often placed in the context of market influence – and we never venture far from consumerist framings – here it is important to reflect on how young people identify and measure other external forces at work that may empower or restrict them. Often there are references to lifestyle, in the Giddensian sense, and society in general in the hypothetical (and fictional) positions they have taken in RGT discourse. However, I also want to draw attention to the struggle, as Gamson has done, between defending the rights of the individual (person or even embryo, if following Catholic doctrine, or even 'deep STS') and the social responsibility towards the many. The opposite to this liberal individualism is of course, consideration for society. 'Parents are one-sided,' said Damien, of NDC. 'They don't consider society.' This is a future of rights between persons, collectives and non-humans.

M1: I think if she has the chance to get rid of the disease you might was well. … You might as well have a full life.

ND3: But if you adopt a baby… You're still having a baby, do you…like. So you might as well have your own.

(ND, *The Gift* film discussions)

The most common utterance of popular wisdom against this rights-based future frame was the ever-present, and ever recognisable: 'Playing God'. Even a secular vision of this, which Fuller (2011) speaks about, is a future design of life, of species.

As seen with our earlier performative methodological approach, gender was a differentiating schema when reflecting on Ryan or Jennifer's life choices. Boys were sometimes unengaged in the perceived female topic of RGTs. Sometimes the girls would goad the boys into discursive action. Yet many males and females who adopted the Patient/Parent frame defended Jennifer's actions, agreeing with the following ND participant's emotive pronouncement: 'It's her baby and it's her body and she should decide what she wants to do with it' (ND, *The Gift* discussions). In stark contrast, Derek – whether or not he was using his aggressive argument approach to performative effect – invoked sexist, derogatory language against Jennifer for not accepting PGD (this was responding to one male's proposed scenario where Jennifer might regret not using PGD, potentially returning passively to the doctor, who might be justified, according to this student, in saying 'I told you so').

This gendered aspect of the frame was a variation on Patient/Parent framings, markedly in opposition to Future Child or Future Humanity framings, discussed shortly. While a Patient/Parent future framing sided with the rights of either the mother, the father, or a patient waiting for treatment with HESC technology rather than foetal life or the wellbeing and identity of future generations, the Sacred Embryo, Future Child, or Future Humanity framings focused on these groups at the expense of the parent undergoing PGD tests or the patient with a life-threatening disease. These frames were coded in this way to make a distinction between how participants used a process of emphasis and selective exclusion for the agency and rights extending to:

1 the immediate recipient of the technology;
2 the embryo;
3 the future child; and
4 future generations or the future of the species.

Each of these four drew on a common sense 'moral obligation' (to use Karen's words) for that frame. In a situation where Jennifer was called on to make a decision about having a baby, there was general agreement across schools that she should 'do something' and that 'something' could include IVF using non-carrier gametes or adoption. In fact for Karen and Des, adoption was the moral solution.[1] The morality of decision-making also surfaced in the Patient/Parent frame in *If*... discussions. Even when Andrew's patient's rights are being invoked, an added dilemma materialises in the narrative when his fight to allow stem cell research is supported by the mother of a very ill young girl with diabetes.

In a changing landscape in Ireland for women, 'social freezing' is a commonplace procedure at Beacon CARE Fertility – the eggs of thirtysomething women who have not met a suitable partner and have used this form of insurance for the future. One wonders to what degree control over women's bodies has transferred from the Catholic Church in Ireland to the pressures of a biosociality of technoscience, embedded in the Patient/Parent future of technology in its Heideggerian essence, a concern of the anthropologists Sarah Franklin, Jeanette Edwards and Marilyn Strathern among others?

A debate ensued in Louth (LH) to my question about degrees of deservedness of treatment. Although many agreed that everyone has equal right to medical treatment, there were others who felt treatment should favour the diabetes patient. A reason emerged for this in LH: the diabetes patient was born with the condition whereas Andrew knew the risk he was taking as an untrained mountaineer. 'Stupidity got Andrew into this situation,' said one. 'He had more control over his condition than she had over her's,' said another. Taking a risk means the risk-taker must take the 'blame', thus suffer the consequences, an example here of how risk experience of this kind is sequestered, taboo, allowing room only for fate to intervene with tragedy. Yet some – Lisa from LH in particular – identified with Andrew not just because of his injury but also the lifestyle associated with it. Thus a person's lifestyle should *not* be considered in preferential practice for life-saving treatment. 'Andrew happens to be adventurous,' Lisa told me in interview. These are classic medical ethics debates which suited the content of *If*...

The spectre of global black markets in organ trading is demonstrated by how the Chechen woman in *If*...was coerced into selling her entire uterus. This opens the discussion up about the vast network of global sociotechnical forces that are extending many degrees of separation from the intended parent who wants to screen, or the patient who wants a cure. What are revealed are nodes, the compositions of application, knowledge, proximity and distance. Young people might use a liberal individualism future projection to remove reproductive and genetic choices from the (structural) regulatory to the personal, as McDonnell and Allison (2006) suggest. Perhaps any ideas of the personal, the proximally relevant, or the spiritual disappear when the economic and scientific realities of the biosciences agenda are set in Ireland. However, there is an argument that it cannot be left to individual choice alone if, for many, this choosing involves an impact on future social and biological identities or future characteristics of the species (as is the case of PGD where embryos are screened for 'disorders'), or, for others, if reproductive choice leads to the destruction of foetal life. The sociological argument made here accepts multiple ways of knowing and appreciation of genetic ties and the embryo, and this is made all the more relevant as Ireland comes into a new stage of development with its many immigrant cultures. Although Jasanoff (2005), among others, hopes for a new politicisation of biotechnology that allows for global democracy, in many senses the biosciences are already politicised, more so as they become increasingly socialised (Nowotny *et al.*, 2001).

'Sacred embryo'

In this section we metaphorically enter the hallowed grounds for a certain minority of students in all schools. Debates that constructed these frames began at times to get heated and required strong facilitation to allow participants to air their views if they wished, and to allow others to do so without intimidation. I must also accept that the deepest expressions of feeling for the issue at hand may have been kept private. As with media debates, the embryo became the central, passionate rights issue, albeit here in a performative sense as participants brought a taboo subject into a formal discursive practice within science class.

There has been a sharp decline in the influence of the Catholic Church in Ireland, perhaps steeper since the main empirical study was conducted with more information now available through various reports on institutional child abuse and the Catholic Church's role. In any case, this decline has been tracked in life political issues of abortion and IVF by McDonnell and Allison (2006).

There is a strong link here between embryo and Harmony with Bifurcated Nature framing, a world where there is certainty of divisions: human essence and non-human. The yuck factor is invoked once more. There is a shoring up of cultural resonance and symbolic packaging is all the more intense to protect this identity. I use 'sacred' here in an academic, secularised sense in that there can be a 'secular sacredness' attributed to a practice or object, something untouchable, deserving of utmost respect, beyond reproach, but can ultimately be either a humanist or religious worldview. I distinguish between three types of sacredness in young people's responses (Murphy, 2008):

1. The embryo is not a person and there is no ***personhood*** until there are human features in foetal life. Phrases emerged such as: 'When it starts to get little eyes and nose and stuff' (Lisa, LH, *If*... film discussion); 'It's not in your arms'; 'But they're not breathing' (to which the counterargument was 'But they're still alive') or 'You're not going to have a funeral for it' (all male, SK, *The Gift* film discussions). These suggest a similarity to other religious perspectives on pregnancy besides the Catholic faith, such as physical continuity, epigenesis, quickening and ensoulment (Maienschein, 2003). Much of modern framing techniques by pro-life movements tend to emphasise the enhanced human form of the foetus as seen in new 3D and 4D scanning techniques. Foetal life is felt, seen; it is *seen life*.
2. The second non-religious, but sacred, category, embryo *is* human, with some kind of spiritual/metaphysical human essence. The sanctity of the body and the social or cultural totemism we have built up around life origins are enough to protect it. Karen from SK and Des from NDC proposed personhood from conception. There are comparisons here with German theoretical 'dignity' framings or Habermas' (2003) inviobility. It is a different kind of secular sacredness – the potentiality of the embryo, *natality*, a word used by Arendt (1958, 1959) to describe the potential, the possibilities of prenatal life, but this offers us a different perspective than the standard pro-life position.
3. In a third category – for a small minority of participants – the embryo was framed in a religious sense as fully human, and thus sacred, from conception. It is the standard Catholic position. In the case of Lorraine from LH, references to the Catholic doctrine were used. Karen from SK might however

have utilised the religious frame also without explicitly mentioning her religious perspective. It seemed to be reflected in her descriptions:

> ... like the making of the world and all this stuff, the scientific view of it, like evolution and then the ah, Adam And Eve [in] religion and how scientists have a different view on life than people who have really deep religion and how you bring the two of them together is really your own... I don't know... experience or whatever.
>
> (Karen, SK4, interview)

> Sometimes...yeah when you think a lot about religion and then ...you kind of have to mature your views on religion. Like the whole Adam and Eve thing like...
>
> (Karen, SK4, interview)

Lisa gave a tentative judgement when interviewed on whether embryos were human or not, needing the justification of a cure for serious disease although she used the understated, sequestered-experience term 'off-putting' if the subject was a 'person'. Other views expressed during the LH debate also showed the 'sanctity' of a fertilised egg for reproduction as distinct from an embryo for therapeutic purposes. Furthermore, there was clear discomfort with using embryos of any kind unless really necessary, for instance resisting their use for cosmetic therapy. It is intriguing that there is an apparent contradiction here – embryos can be used to help sick people if they are not going to be implanted but not necessarily for cosmetic reasons. By drawing on embodied features to describe the personhood of the embryo as Lisa does, she draws a boundary to human sacredness. Yet she has not been convinced that the embryo should be used for all purposes. This collective identifying of a strong position followed by more tentative musing afterwards on using embryos was also evident in SD, a school that was strongly pro-technology. The mechanism of classroom debate is reported in more detail in Murphy (2008) regarding these two students, but for Karen (SK girl) and Des (NDC boy) they stood their ground as lone (public) defenders of the embryo on religious or quasi-religious grounds.

'Potential' was a word used by participants in SK and NDC with different frames to describe the personhood of the embryo, also recalling Arendt's 'natality.' While Damien from NDC saw 'potential' as effectively 'humanness', his classmate Derek and other respondents from SK used potentiality as pre-human, thus non-human. Potentiality for Des is enough to be considered human. To him, a cell can in a way be human, or part of human essence. He gives no distinction between an embryo in a lab, cloned or otherwise, and an embryo in the uterus destined to be a person. His reasoning for sanctification at fertilisation was because of the 'vagueness' of an arbitrary scientific decision on when life begins (not unlike Jasanoff's criticism of Warnock (Jasanoff, 2005, citing Warnock, 1985)). But it remains to some extent a quasi-religious description of the embryo: it must get 'spirit' at some stage along its development. His wording is interesting, bringing to mind the contingency of human constructs: conception is the best 'choice' for 'deciding' on personhood because anything else would be too 'vague'. The act of 'screening' of embryos is interpreted as 'testing' only for informational purposes. This is acceptable – the act of then getting rid of the embryo if there is a positive result for disease is not acceptable. The core humanising frame of Des becomes shorn of argumentative strategies or symbolic packaging when he finally draws on the emotive image of media pro-life framing – embryo selection is abortion and abortion is murder.

Religious frames often drew on particular mythic themes. The cultural theory word on this goes to Damien, who described religion stories such as Adam and Eve as a method of control through metaphor, a metaphor which humans have 'dug themselves into a hole' by using specific explanations through stories. The prominence of origin myths can be traced across the many conversations about conception or birth itself (like Rose, 2003). And as Damien says, 'origins need a creator'. While Lorraine draws on the biblical term 'creator', Damien does not rule a deity out of this analysis, with his word for God being analogous to risk rather than Fortuna.

> Because like y'know the universe was there always. Like there had to be something that started, I think. Whether you call that God or chance or whatever. But there definitely was something in the beginning.
>
> (Damien, NDC3, interview)

'Future child'

Another rights-based future framing concerns the offspring directly, that is the progeny as child, rather than some current potentiality as 'embryo', 'zygote' or 'blastocyst'. The rights frames can often be set up Future Child in opposition to Patient Rights; there are the anxieties generated by risk calculations themselves, plus the problem of screening out 'unlikely' contingencies, thus reducing life-planning to manageable proportions (Giddens, 1991: 182). In *The Gift,* the risk of the future child Mark getting FRDA was one in four, which is statistically high, and therefore more than an 'unlikely contingency.' The UK HFA act enshrined the 'welfare of the child' above anything else. A UNESCO report on PGD from 2003 states, following a meeting of a working group of leading experts, that 'the issue was raised whether a child's "open future" is sacrificed through an uncompromising respect for parental liberty in reproductive decisions, including avoidance of potential harm' (UNESCO, 2003: 13).

In media controversies, as we said, framing reduces orders by lesser complexity, but raises the issue to the public or political. These reflections and agreed understandings of world structures as represented through media managed to inspire in many a linking of events within the issue culture of RGTs, and this in turn provided the basis of a type of activism from their positions towards school, church, or state. Laura from LH was critical of the Catholic Church's teaching about embryo research and related technologies and emphasising the Catholic ethos of her school was not an influence on her views. For the purposes of my study, the cross-linkages students made themselves between cultural references and RGTs were sometimes weak, but the desire to actively find out more, and contribute to its change through knowledge, was strong. One participant in particular joined the general rebellious nature of this school's participants (covered in Chapter 8) by exclaiming she resented any government input into decision-making on IVF/HESC regulation and matters around when life begins.

This critical streak links many of the schools, from Deirdre (ND1), Katherine (SK2) and Kate's (SK3) aversion to stereotyping of their age and subcultures, to Lorraine (LH) who, though a practising Catholic, recognises the importance of having an opinion in religion class. Youth is the age of rebellion and it is a necessary part of identity-formation. Statement 2 on Ryan's right to information about his genetic make-up (see Appendix) became the most unanimously agreed statement throughout all schools: Ryan has a right to know he is carrying the gene for FRDA. There is no doubt this 'youth power' aspect of the film caught the imagination, particularly in ND.

The responses suggested a life-planning approach, in the midst of disease. By 'knowing' Ryan had the 'gene' or the' disease', he could plan his future. Fortuna would appear to be missing from the arguments. Or is it? Are there traces of determinism and fatalism mixed with Modern Progress and risk within the logic that he has the gene, he will get the disease, he needs to deal with it? The high chances of getting the disease translate into inevitability, fate. There are also suggestions that when a severe physical disability appears to be imminent, your life chances are diminished. This is not, however, a traditionalism that leaves people to their places and accepts fate. This is autonomy for young people, a life politics that empowers them to change it. The phrase 'full life' recurs again and again. One NDC participant used the embodied Jennifer argument: Ryan has a right to know about his body. Knowing this information is a further part of knowing your identity. A very small minority in schools preferred not to know at fourteen, although there were some disputes about what age a person can deal with this life-changing information. A general consensus left it at various levels of maturity, rather than a specific age. Young people feel distanced from decisions about predictive genetic testing. It is the emerging confidence of a sixteen or seventeen year old that causes them to want to know, have a *right* to know everything; and perhaps, to some extent, they feel they do. As one ND participant observes: '… at sixteen, you know what you're going to do with your life and anyway' (ND, *The Gift* discussions).

Most who framed the debate in this way recalled to the discussion how Annie suffered and, for many, it would be inhumane to allow this suffering to carry on to the planned child. There was however a counterargument position in ND that framed the future child another way: it is wrong for parents to control the fate of their child. PGD here becomes, as Habermas (2003) says, 'an invasion in the deontologically protected core of a human person' (ibid. 87). This was a solid position in debates, partly influenced no doubt by Mark's rejection of his genetic make-up at the end of *The Gift* to pursue the life he wanted without any more interference from his parents. PGD becomes a threat to individual identity: 'He isn't [to him] an individual' or 'They're meant to be who they want themselves' (ND, *The Gift* discussions).

To the statements in *The Gift* activity that refer to the welfare of the child, participants offered this criticism to a parent using PGD: 'It's what *you* want. It's not what *the child* wants' (ND, *The Gift* discussions). Although possibly unaware of any wider nature/nurture discourse, another respondent in ND explains: 'It's how you bring up the child wins over.' Many talked about the selfishness of the parents, denying the right to life of a child with disability. It was difficult not to see in ND responses glimpses into their home worlds, whether it was mimicking their parents, references to parental conditioning, or a crisis in their lives generally sequestered from biology class.

And so it is not, in this framing, just a grand idea about genetic determinism, technological progress, or science's control over humanity that is the problem. It is human agency controlling human life. And for some there is a strong urge to leave this to Fortuna/Fate. 'He's making his child,' an ND girl says. 'He's not letting his child be made. He's doing it.' This sounds very like the classic grown/made distinction (ibid.)

What often was difficult to do in these debates was establish a boundary for types of disease that should be screened. This is where the contentious area of eugenics rears its head. For those who look at future society as a whole rather than this one pre-selected child, the concept unfolds out to a broader social programming canvas, and here we can tread on yet more dangerous social ground. A large majority across schools drew a line at life-threatening disease only. Yet when other conditions are mentioned:

> Like if you're going to see your child bullied for the rest of their life…or being discriminated against… I don't think you'd want that for your child.
>
> (ND, *The Gift* film discussions)

'Future humanity'

This is a rights-based framed future looking at anti-presentism. These types of future outlook suggest an agency either for 'humans' that exist now or will at some stage in the future. Relations are seen here in terms of power, rights, commodity, and human recklessness. Like consideration for Future Child, the marked difference that RGTs now bring is the possibility of one generation planning the existential identity of the next to the extent that that future generation cannot say they own their genes (Campbell in Esposito, 2008, translator's note, x; Habermas, 2003). It is a true colonisation of the future (Giddens, 1991).

The concept of rights, of course, was introduced in many statements called out in film discussions. As stated already, there was the 'right to argue', but also

(Young) Ryan's right to be tested and (Older) Ryan's right as a father to give his child 'the best'. Yet I allowed students to make what they wished of the democratic and libertarian nature of the statements. The first statement in *The Gift* activity is perhaps less obviously revealing a democratic slant: 'Jennifer is right to argue that even though she knows that there is a risk of her child inheriting Friedreich's ataxia, she wants "nature" to take its course'.[2] Students might interpret this in two possible ways: what should Jennifer do that is for the good, or is it for the good that Jennifer should argue her case, regardless of the final decision? My own research framing of the question as socioscientific deliberation might well suggest to young people that they must now ponder on what they think is right or wrong, so the former interpretation is taken. Perhaps once in each session, a student might suddenly see the latter interpretation and cause the other participants to pause and reflect on what the statement is asking.

> Have a natural born child like, she's right to argue it, not saying that she should do it.
>
> (ND, *The Gift* film discussions)

NDC1: (Derek)
Yeah but she has a right to a child that [inaud]
NDC3: (Damien)
The question is the right to argue though, not lead to a child...
All: [raised arguments]
NDC1: If you can't lead your life fully though.
NDC3: That's not the question. The question is the right to argue though.
(NDC, *The Gift* film discussions)

> Is it not asking like her right? Like if I was doing it... if it was talking about myself I would say that I should get it screened but it's her right...
>
> (Kate, SK3, *The Gift* film discussions)

As a riposte to Kate above, her classmates were quick to invoke the former interpretation again, implying that perhaps, Jennifer does not have the right to argue if there is no moral obligation to do so. The counterargument produced in this example appears to have a gendered element:

F10: But does she not have the right to listen to her husband as well? Her husband's ... knows what...
PM: Yeah.
F10: ...will happen like. It's not just her decision.
F11: Yeah that's true.
PM: OK so but...she had a right to argue it but you're saying she

F10: but she still has...she still has to listen to her husband because he knows what happened already.

PM: Yeah

F10: And he knows what the mother will go through with the child...

(SK, *The Gift* film discussions)

Both Future Child and Future Humanity framings demonstrate a *motif* of responsibility towards society, although diagonally opposite in some respects and might push the limits of 'practices of the good' in their extreme application by the criteria of most participants: a eugenicist might be in the Future Humanity universe while a narcissistic parent might find themselves in the Future Child one. It is beyond our scope here to further seek out degrees of communitarianism or deontology or socialism in responses. However the emergence of a politically inclined caring for others in a world village would not be overstretching the mark in these concerns. Contradictions though, as ever, remain: in beginning the process of PGD, although it may have been initiated to remove an embryo that would have lead to untold suffering of a future child, how can there be an argument for random selection of remaining embryos thereafter (as Jennifer wanted in *The Gift)* as opposed to more selective procedures without reintroducing Fortuna having already begun to 'play God'? And how could these four centres of power in discourse (Patient/Parent, Sacred Embryo, Future Child, Future Humanity) resolve their differences without the almost whimsical Parliament of Things devised by 'practice theorist' Bruno Latour (2004a)?

In addition to linking with Fortuna and risk, either of the autonomy/rights frames could align with Modern Progress or Harmony with Bifurcated Nature. For example, scientific rationalism could support Ryan's right to rid the next generation of Friedreich's ataxia or more democratically, the future generations' right to be born free of the disease. It could also, as Damien did, support Jennifer's decision not to use a form of RGT on the grounds that such technological meddling could have a detrimental effect on natural selection. Likewise there might be a scientific basis for the personhood of the embryo offered, just as there could be an argument that the embryo is defenceless against the onslaught of either technological individualism or a collective social acceptance of a technology. A few participants highlighted the danger of removing 'a gene' without fully knowing the consequences. What if this same 'gene' also had a beneficial effect on some other biological process?

I give the last word again to Deirdre from ND, who, as already stated, had experience of RGT issues in her family:

ND1: My sister said that if she had a chance to go back and change Millie she said she wouldn't, but to have another child ... bring another child into the world with the chance of it having a cleft lip and palate she wouldn't do it.

'Commodification'

For those young people who placed PGD into a Commodification future framing however, there appeared to be no such direct link with Modern Progress or Harmony with Bifurcated Nature. This was the process of using product- or lifestyle-orientated descriptions for embryos or future children to frame RGTs linked with money or broader neo-liberalist concepts. There were linguistic clues here to negative framing of the technologies as participants spoke of 'shopping for eye colour', 'shopping for children', 'whatever child you want you can have', 'picking characteristics', 'picking out…blonde hair, blue eyes', 'he was *used*', 'the perfect child' and so on. The effect of increased reproductive choice, as Strathern (1996) was aware, is increased exposure to market forces. Whereas *If…* dealt with commercial interests as a major theme of the human egg/womb trafficking industry, *The Gift* had just very slight reference to a Commodification society when Mark says to his father Ryan near the end 'I'm just another dream your money can buy' (*The Gift*, 1999).

I looked here for Giddens' (1991) lifestyles of authenticity still emerging from de Certeau's (1984) 'make do' merchants – young adults who were knowledgeable of the market forces around them and who drew practical boundaries that were not exclusive to commerce in these life political matters. Lisa provided such a view. I already alluded to Lisa's identification with 'adventurous Andrew' in *If…* and her reluctance to blame him for his mountaineering accident because he knew the risk involved. She supported his quest for a cure using HESC. Her classmate Laura, too, thought hard about not being able to take part in active sports again. For Lisa, confidence was the most important attribute. Hers may have been a Celtic Tiger view, of its time, infused with the pragmatism of consumer needs. We spoke about the TV show *The Swan* where individuals are brought on as 'ugly ducklings' to undergo plastic surgery and redeem themselves as beauty queens at the end of the programme. Lisa loves this show. We have already mentioned her acceptance of Alex's possible motivations of financial or publicity reward for scientific endeavours. However to simplistically label Lisa as following a pattern of neo-liberal capitalism for embodied improvement is to ignore Giddens' lifestyle politics: 'a time/space "slice" of an individual's overall activities, within which a fairly consistent set of social practices is followed' (Giddens, 1991: 243). Mediated culture's representations of confidence and appearance is taken in to her lifeworld with many other representations. We must though, with Giddens, accept the current central role of the capitalist institutions in our society. Yet it is the authenticity of the individual's reaction to consumerism that counts, and this may be for another discussion. Inevitably, in mid-2000s Ireland, practices of consumerism were constructed through the languages of consumerism. And arguably they still are.

An intriguingly common recurrent phrase from young people was 'everyone will be the same'. This was how many young people saw the natural development of the human race with lightly regulated RGTs and the urge to buy perfection. Habermas' 'dedifferentiation' (2003: 44), a new shaking up of 'grown and the made' (ibid. 42) would appear to resemble this popular wisdom. There was a visual element here of how the natural endpoint of the combination of human desires, rampant consumerism, and new technologies would lead to a generic engineering of the human race, summed up by Wanda from SW as a race with 'no failings'. The increasing amount of plastic surgery among celebrities might demonstrate this moulding similarity, bringing further existential danger from totalisation of technology in a Heideggerian way (Campbell, 2011: 7). There was a strong feeling that things would melt into one, threatening diversity.

Individual or collective parental choices might affect our pure relationships (Giddens, 1991), sacred familial or other bonds which the expert systems have certain control over through PGD or scientific research. Commodification, autonomy tied to consumerism might in the counter-narrative of in this world, ultimately lead to authoritarianism, allowing in horrors from our past:

> Like…as I was saying it evolves… Science always is evolving every single day. And if they do find out the genetic code – the code for the genetics like – they could actually like… but then again it's like shopping for your baby. 'I want my baby to….' It's like fascis[m] or Hitler. German people are portrayed as blond hair, blue eyes… which is really just stupid. So if it comes to that in the future…will there be another dictator? Will he make everybody have blond hair blue eyes? Like so on… like that's a whole other race of people. So I think that's kind of scary.
>
> (Deirdre, ND1, interview)

'Pandora's slippery runaway'

This framing is a cod-mix of Pandora's Box, the 'slippery slope' argument and the 'runaway' frame from Gamson and Modigliani (1989), things getting out of control. There was a feeling among many that the technology would run out of control, that humanity cannot be trusted with its stewardship. This was not necessarily an opposition to scientific rationalism but, again, to humanity's technological applications, a fear of its innate prejudices and selfishness. Dystopian scenarios of a genetic inequality like *Gattaca* (1998) can, as we have already said, present a piece of jeremiad rhetoric, but Hollywood sci-fi can go much further into nightmare imaginaries.

> 'There's always somebody who goes to extreme.'
> 'You don't know where to stop.'
>
> (Deirdre, ND, interview)

> 'The time limit on stem cell research [on embryos] will keep on increasing.'
>
> (Lorraine, LH1, interview)

P: Once you start doing that kind of thing you be start picking out 'Oh well I want my child with blond hair and blue eyes' you know what I mean?
L: Aw not like that.
P: If you go that route...
O: It wasn't like that like that...It's just a disease we're talking about
M1: Yeah but that wasn't the question.
P: It always happens like that. If you do one way ...a less thing,
N: Yeah you don't know where to stop
P: ...there's always somebody who goes to the extreme with everything.

(ND, *The Gift* film discussions)

Interesting sci-fi type of future scenarios arose. Although few participants I asked had seen *Gattaca* (1997) or read *Brave New World* (Huxley, 1994 [1932]), the landscape of rich 'perfect' people ('super-people' according to an NDC contributor) and a genetically inferior underclass was painted. Recurrent phrases like 'looking for perfection' or 'the perfect child' appeal to how humanity's baser instincts would abuse science. Damien's references to science as a structure – 'humanity hasn't learned to respect science' – or the popular wisdom of 'stood on the shoulders of [other] people' might have elements of the very 'God replacement' he critiqued in our earlier account. Here, science is independent of human agency. Damien saw a natural evolutionary change for the better might occur without genetic engineering.

> I just think that ah... we... we've gone so fast with all the technology we had like... we've kind of just stood on the shoulders of the people that did, we didn't really kind of earn it ourselves. So I think because of that we haven't really earned the respect needed for such technology as that. We kind of just... we don't really see the consequences of using this technology. We just see it just as a new kind of invention, let's give it a go. I think there's much deeper consequences of it than is first seen.
>
> (Damien, NDC, interview)

All who entered the world of Pandora's Slippery Runaway with a critical edge wanted clear boundaries or else it was a runaway world. On sex-selection options, very few students in any school could give justifications for this. China's birth policy was presented as an example. Although Kevin from SK trusted the systems of governance to regulate ('Anyone can't just walk in and pick their child like... there has to be some sort of ... Way of doing it.' (Kevin, SK, interview), dystopian imagery darkened many reflections, which would certainly have been shaped by both films' premises, and sci-fi generally. Deirdre from ND articulates young people's fears best:

> Anything could happen really. Like there's not really anything that couldn't happen. The things that could... like to read the genetic code... I don't think

many people would be able to do that. Cos I think they'd want to keep it kinda secret. Like all Governments do! (laughs)

(Deirdre, ND1, interview)

Notes

1 In classrooms, we did not have recourse to moral philosophy to investigate how Kohlberg (1981) or Beauchamp and Childress (1994) might have tackled these dilemmas. I do not have a background in applied ethics or moral philosophy, from which so many media 'bioethical debates' draw their store of knowledge. All we, as a discussion group, had to draw from was common sense and some way of following a route of practices that leads to the good life (Habermas, 2003). This good life often took the flavour of what anyone might do in any given situation, taking on their normative practice solutions rather than formal moral philosophical arguments. This is key to my argument about exploring practices in education, both for content and reflexively for pedagogy.

2 This first statement in *The Gift* activity is rather ambiguous and caused some confusion in schools. What did it mean to 'do something?' By challenging participants to ask themselves would they carry out an alternative action to Ryan to prevent Friedreich's ataxia in an offspring, as opposed to 'doing nothing' *except continuing with your natural plans to have a baby,* I was opening up to more 'positive options', to use a phrase borrowed from anti-abortion discourse. Initiating a response here might have contributed to the poor support for Jennifer's wish to have the baby naturally, as referenced below.

Chapter 8

Life politics

> He didn't want to eat a pigoon, because he thought of pigoons as creatures like himself. Neither he nor they had a lot of say in what was going on.
>
> *Oryx and Crake*, on young Jimmy (Atwood, 2003: 27)

We have now covered many aspects of RGTs in science education from the standpoints of ideas and practices that emerge through the gates of the school from biopolitical Big Science, news media, governance, prospective parents' decision-making, even science fiction. In the quote above, Jimmy, the son of genetic engineers in *Oryx and Crake* (ibid.) identifies with the pigs his father genetically engineered to carry extra organs for human xeno-transplantation, called 'pigoons'. Both neglected son and pigoon are caught in structural compositions – bio-highways of action and agency in which they have no control. The quote could just as easily have been about a consumer and a factory chicken, or indeed a stem cell patient or a 'uterus donor'. Life politics is the politics of decisions and the earlier we learn this, the better. In this final chapter, I shall conclude by bringing education into politics, politics having been already (re)introduced to biology education. The argumentation of young people serves them well as critical thinkers in a different identity-oriented politics of the future. We will look at how presentism can be reduced in these arguments while introducing reciprocity rather than social contract. The result is a more engaging biology pedagogy, and an engaged bioscience as part of future governance. The Parliament of Things begins for the biosciences and their networks. Small changes in approach to second level education will do this, and the chapter concludes with practical advice to teachers.

Let us bring together the main themes from previous chapters. There is a challenging question that links contemporary 'Mode-2' science education to future biotech governance, acknowledging that it is important to include the personal and ethical dimensions of RGTs as well as speculative reasoning on future populations and humanity: how do we negotiate a future constitution, as Bruno Latour rhetorically positions it, for mutual temporal (intergenerational) communication of modern genetics and its societal implications, starting from secondary education? Life politics in the book up to now will have been defined as the type of politicking involved in various public spaces that utilises the traditional tools of politics and rhetoric mixed with new imaginings of biopolitics (governance of biotechnology, genes, molecules, etc.) as well self-actualisation of individuals and collectives (the cultural studies side of this, first realised by Anthony Giddens (1991), as expressed in particular by young people in a biology classroom).

Biopolitics would suggest humans and non-human agents are trapped, to a degree, in common lattices of action. Non-debated and unanswered genetic screening can go this way – a 'Euro-American' (Strathern, 1996) culture that aspires to beauty and eternal youth but of the moment and free of disease. But every so often, there is a pause, an exit from standard normal practices and an entering into another discursive realm, one that might be more public, more formal, more audience-orientated, more dramaturgically performed. This is life politics. There is the means here to give young people their real autonomy, their identity and self-realisation (Giddens, 1991), not the self-absorption of current youth discourse, not *kynicism* (Sloterdijk, 2011) but tied to the external referent of future bodies and future governance. Will bioscience and choice in an era of neo-liberalism lead beneficiaries to be, as so many young contributors in the study said, 'all the same'? Will we use RGTs as a 'supermarket'? So we are within practices, but there are limits to what action and analysis can be done through observation of practice alone – humanity, the transcendent, the good life, the right thing to do. How do we do the right thing? Once broken free of presentism, can it be for the common good in a posthuman future? How do we get control of our biology with common purpose, with sets of criteria and guidelines? How can we see these in practices?

This final, overarching conclusion is a concept that reaches from classroom to future governance. Suddenly science pedagogy, using embodied movement and media becomes an event. This chapter will discuss how future generations need a type of trusteeship with publics, while applying morality of the everyday to values in science as it would apply to human genetics.

In the intellectual context of STS and science communication, there is a move – although not a fully shared view – to a democratisation of knowledge. Life politics can be a contemporary oratory-mediated argument of the changing concepts of life in structure and practice. RGTs are completely caught up in this. The education system needs to respond. Giddens and Latour are the intellectual touchstones; but within this democratisation of knowledge, where are we with regard to future human life (a species and subject of human question) when this democratisation extends down to molecules and genes, up to species, biota, states, biopolitics and stars, and across to bioinformatics and media? How do young people save the 'human' and what of this needs to be saved in any event? How do we make this education systems part of reciprocal future governance, an education that is also part political forum for biology?

Bioscience and choice

We will have seen through the course of this book how the emergence of RGTs presents us with immense opportunities and challenges for future parents and society. It is not easy to see the emergence of a social contract – and we must also guard against a contractarianism – but it was a Rousseau ideal between society and state, between individuals and regulation, between the person and the markets. Whatever about the markets, for science it would indeed be a light contract, where a Leviathan calls science to account fully for the people, and not a phenomenon that could ever have occurred practically in any case given what we know in STS about materiality and biosociality. The alternative to a Rousseau/Hobbes social contract, at the other end of the spectrum of scientific autonomy, are guides and shapers of RGTs emerging from industry subjected only to the laws of the markets. Giddens' Third Way, as controversial as this sounds to many on both sides of the spectrum, can be the best alternative, structure and agency constantly interacting.

Justification requires argument from different regimes of everyday personal and social practice (Boltanski and Thevenot, 1991). Giddens tells us those personal/social lines are blurring, and when you see how fame and infamy goes viral for one person on the internet, you see the evidence. Even the 'new spirit of capitalism' (Boltanski and Chiapello, 1999) has constrained us, but is now out of date with recent market collapses and other existential crises (Giddens, 2009).

The limits of practice are clear. We cannot always be trapped in bodies of habit, neo-liberalist order, essentialism, unsustainable consumption patterns, individualist core tendencies, without some innovation; though that cannot get to its ideal centre, its self-actualisation without risk of becoming a 'morally stunted process' (Giddens, 1991: 9). In Facebook, tweeting, friending, new communities are being constructed online. The corporate world of global media owns us, so how do we use tactics to resist that, to take back our data, that which is, increasingly *us* – genes, codes, bytes? Life politics challenges us to look at internal practices, but always by means of having an external referent. This does not have to be complete Kantian transcendence, but there is a type of transcendence in the immanent head of steam built up by everyday practices. In other words, value accumulates, and the value of the human, or the object, or the relation can remain, and intensify.

Habermas, in quoting Andreas Kuhlmann's *Politics de Leben*, says: 'Parents have always been given to wishful thinking as to what is going to become of their offspring. Still, this is different from children being confronted with prefabricated visions which, all in all, they owe their existence to' (Habermas, 2003: 53, quoting Kuhlmann, 2001: 17). The walls between young people – and the students in this book as they get older – in how they confront an opposition view on RGTs are clear, even if the way that walls and windows can be created for governance and regulation is not so clear. To take the example of Down's Syndrome (which SKt told me at the end of our work, his child suffers from) – we could morally argue that parents need not have a Down's Syndrome child, while feeling privileged that such children currently exist. How do we do this? Just as we need to use and confront seen and unseen life, so there is also *said* and *unsaid* life. We need to talk with care about the form of those who now exist who may no longer in future (or be dragged into Heidegger's breach of man from human (Campbell, 2011)). And of course, as we speak of frames, framing a pregnancy will depend on circumstances, expectations and current requirements; is it a baby or an intolerable, medical issue, a life disruption? Life and non-life, seen and unseen life, interchangeable, but honest, all 'pro-life' and with its own transcendence.

Future thinking is key. The future is still the past – conservatism of the Euro-Catholic as described in Chapter 2 runs up against Strathern's Euro-American, loaded with choices in a free market (Strathern, 1996) – in fact as Strathern so clearly articulates, both modes of action mix well together. And if the axis of power is complicated, should we deny the right to offer the best for our children or should we think about species morality? Then there is also an assumption that any of us, even the sovereign or the regulators or indeed the markets, can both predict and control the future. As Kelly says, taking Derrida into this:

> On both sides of debates over genetic engineering and cloning, we see that philosophers assume a sovereign liberal notion of the individual who is free to choose, who can make decisions, and control the future....This is not a question of whether the state of current science gives us the knowledge necessary for such mastery and control but rather the question of whether or not it is possible to possess it at all, ever.
>
> (Kelly, 2011: 42)

Haraway (1997: 265) says in her darkly humorous postscript: 'Ties through blood including blood recast in the coin of genes and information ... have been bloody enough already'.

But while both Latour and Giddens acknowledge a politics (of Nature for the former, of Life for the latter), neither brings us to the practical of what we will actually do in this assembly in this agora. Persuasion and rhetoric will be needed after all (contra Latour and Giddens), but so will the personal-political. The best arguments will win – they always have.

Public opinion or popular wisdom?

In Latour's compositions of phenomena, we see opinion/fact complexities, where fiction meets matters of concern, and within these 'factishes' are co-constructed mediated material from the labs (the core-set of practical knowledge (Collins and Evans, 2002)), out to mediated co-construction, out to ideas and folklore taken from what Gamson (1992) calls popular wisdom. Importantly though the biosciences cannot be said to *start* from the core-sets of lab expertise. For Latour, cultural popular wisdom is also a starting point of these complexities. As we saw in

Chapter 6, there are idioms from both child and scholar which might fit this category – on the cultural side of genetics race, blood and parenthood: 'everybody will end up being the same', 'slippery slope', 'everybody knows' (these are young people's tropes, as found in my schools research). I will summarise here on the limits of using public-constructing opinion as a branch of government and governance, often simplified for media without nuance the 'phantom public' free of thought communities or core-sets of expertise. Why is this so important to the future of human genetics? The potential impact on education and public pedagogy is that both science teachers and science journalists act as mediators rather than intermediaries (Latour, 1993). Compositions (as literary as it sounds) or complexes (with its synonym of technical hang-ups among teenagers) require relations between 'things', where we focus, but also some element of co-production (Jasanoff, 2005) done in a convivial, tasteful way (Boisvert, 2010), with *eloquence* (Latour, 2005b). The elegant solution to a horrific life-changing bioethical decision, an eloquent summary, or a theory – they are all also pragmatic.

The brashness and coarseness of mediation will persist – how could it not? – for instance, as a (mediated) positioning strategy only, as we said in the first chapter, it infantilises, necessarily over-simplifies – 'hard, cold logical scientific liberalism *versus* fundamentalist, God-fearing, pro-lifers', bringing out the artifice, the mediated, necessary fakery of debate and opinion. Within mediation are the strategies of agenda-setting and the tactics of response and action. The influence of bio-orators remains, yet as we have seen, there are complexities to 'biopinion' outside of core expertise. Governance needs to listen to the wisdom of certain crowds. And we will all need the transferrable skills among young people of identifying and constructing verifiable knowledge amongst the ever-increasing deluge of facts and things, or *dings* (Latour, 2005b).

For the bio-orators of yesterday, the science of life was a bounded subject, and the rhetoric to defend was needed to understand Nature in the sciences, particularly physics (early twentieth century) and biochemistry. For tomorrow's bio-orators, this boundedness can no longer be the case.

Ethics or practical good?

This section will wrap up, with many strings still needing to be tied, morality and the everyday, values in science as it would apply to human genetics. In class, there were sometimes informal discourse, banter, slagging, tacit performance of not-clever and often dangerous talk, a moralising of the everyday, and potential future everydays where various RGTs would be widely available for prospective parents and couples – a stark realisation of terrible reality. We are looking, in ways, beyond ethics. There is a network between genetic testing of embryos, the sampling of animals beforehand, and then, as we explored the themes of *If....Cloning Could Cure Us* in Chapter 4, the connections that force us to confront global, socio-technical ethics. From Nietzsche to Haraway, as Esposito (2008: 46) points out, the immunity – a dangerous necessary evil injected to the body politic – was a trope in studies of social order. Latour's assembly, the Parliament of Things, has such a mechanism (2004a) – Latour's bicameral Parliament of Things contains a Lower House of identifying and organising of compositions and hybrids, but the Upper House brings to account with the big picture in frame ('scenarization of the whole' (ibid.)), through argumentation at the gate of the assembly, so there will always be exclusion, except tomorrow is another day in parliament, and today's outcasts become tomorrow's parliamentarians. There is constant flux between organising an assembly of things and ideas around 'what is disease' and 'what is normal'. As Canguilhem argued: 'An anomaly is not an abnormality. Diversity does not signify sickness' (Turney, 1998: e1, quoting Canguilhem, 1979 [1945]).

A crucial argument for the whole book is brilliantly stated by Campbell:

> What practices might we develop, what kinds of discourses are available to us that make it more difficult to reproduce the immunitary and hence negative biopolitical inflection of modernity? As Esposito argues, no more important question requires a response from us than this one.
>
> (Campbell, 2011: 6)

In a similar vein, this immunitary process should not be beaten off, sequestered, from a science education system that is itself immunised. There is significant research from our scholars that suggests a new humanity or something altogether different is beginning to happen. Adopting and developing RGTs now offers these new threats and responsibilities as humans, as families, as societies, as future generations. To strive to keep those human traces is surely the noblest thing, of which there are so many, for without them where is the question coming from? While Fukuyama (2002) and futurists predict the end of everything, some social scientists since Heidegger could claim we have already began the big move towards an uncertain, posthuman future.

Irish education is a changing landscape, and although there are Catholic traces in secondary schools that surface in the discursive actions of the participants in school sessions, I did not find any evidence of an exclusionary bias based on, or references to, Protestant gate-keeping (see Chapter 2 and Krishna, 1992).

I will however close this section by recalling a school session which took place on a Friday morning in January 2005 and which represents for me in snapshot the possibilities of sharing educational spaces outside the school curriculum for a type of life politics pedagogy. It was a collaboration of science and religion class in one of the research schools. What was key to the success of this was that both teachers and students reconstructed this 'biology class' with a deliberatively open space for exchange of ideas. The combined class contained Transition Year, third year science and fifth year religion class.

Science teachers had informed me that students had been covering abortion in religion class and the teacher was uncomfortable answering the scientific questions that were emerging about the embryo. The teachers themselves were upfront about their own limitations within both knowledge domains and saw this session as an opportunity for overlap. The religion teacher explained to me her nervousness about over-emphasising or highlighting a particular ideology on abortion to her class for fear that it might reveal any subjectivity. As part of the pilot for the introduction of a new religious education syllabus at the time, students and teachers were given autonomy by the Department of Education and Science to construct, with the teacher, their own learning plan from a wide syllabus. The religion teacher presented a choice of topics for the students to choose from at the beginning of the year, one of which was abortion. Most students had selected abortion as top of their list and this was the topic they were covering at this point. The NICB researcher and I combined to present a joint presentation.

The teacher was present at the combined class, contributing some expertise to the scientific elements of the presentation rather than ethical elements. Other teachers sat in this large, cold but crowded room in anticipation beforehand. There were some terms from the scientist's world which needed explaining, like 'published science'. Students were not inhibited from interrupting and asking questions in an information sharing pedagogy. During the talk, a pivotal moment occurred. A student asked:

'How do they decide what is abortion?'

The question came out of the blue. She had a follow-up question: 'From a moral point of view what is the difference between stem cell research, IVF and abortion?' This changed the nature of the discussion from the realm of the scientific to that of a shared value that was ethical in nature. Here at last students got to ask the question they were asking their religion teacher. There could be no satisfactory answer from us – or from me as the 'ethics expert' – on this, except to refer to the issue as a question that is currently occupying ethicists, scientists, theologians, policy-makers, and sociologists and will become a sharper focus in Ireland over the coming years as the Irish Government, at last, responds to the CAHR (Commission for Assisted Human Reproduction, 2005) report. (As of writing in 2013, it still hasn't.)

After more straightforward Q&A, the next phase of the proceedings was also a unique, unscheduled departure from other schools. Most students and all teachers left for their break, or to prepare for the next class. However eight students remained. They included the participant interviewees, who, crucially, were not being interviewed today. These students, the speaker and I assumed, were students most motivated to engage more and ask further questions. I was surprised to see two of the teacher's identified 'troublemakers' there. These students were not, to my knowledge, directly asked to remain for this session. At first they spoke about the session, how they enjoyed it, what parts they liked. They were talkative; even those interviewees who were usually quiet. Then criticisms slowly began to emerge about the school. One complaint followed another. Funds go towards teachers, never the students. It would not be possible for outsiders to see beneath the surface. What the outsider sees is just 'cosmetics'. Why do you think it is so cold in here? They never turn the heating on. Student voice is unrepresented.

A teacher sits on the student council and can override decisions made by students. They want to be consulted on the content of the curriculum.

This was powerful – and dangerous – material for qualitative researchers. Could it be that in this activity they saw a forum to air their grievances which had previously been denied them? The fact that respect, a degree of authority and trust was handed to these eight students by us outsiders may have inspired them to use this to speak to us, the outsiders.

The bioscientist speaker from DCU and I defended the teacher. We felt compelled to explain the teacher's generous time for us and his drive for a more relevant, practical, and society-oriented classroom. Were it not for the teacher's commitment, we protested, we would not be there in the first place. There were moments when students had to watch their guard as their teacher was in close proximity. For the sake of their confidentiality, we could not raise this event in any concrete way with this teacher.

Care needs to be taken to distinguish between genuine grievances and unwarranted attacks on teachers. Students will perhaps always rebel against their teachers and against a structure or system. However I was encouraged by the students' own analysis of the situation and their own quest for solutions. During the three phases of the session a range of topics showed a very reflexive agency among these young people: a lack of trust in Government regulations on RGTs post-CAHR; the need for activism among legal experts who can do more practically than philosophers can; the political power over RGTs should be with the people not the legislature. In many ways, this was a type of life political movement within their school. When outlining the theoretical and methodological development in Chapter 3, I mentioned the critical aspect of inquiry that appeared in my research. This session was the defining moment. This was the moment where I realised I needed to look at shifting power concepts and co-construction of values in educational practices – not necessarily information, or even ethics – that divides normative school science from life politics.

Humanity: so what is the human and what can she be?

> Of bodies changed to other forms I tell...
>
> (First line of *Metamorphoses* (Ovid (1986: 1))

Should we care that we head towards the transhuman, the already *post*human that, paraphrasing Whitehead, the present subjects of genes/embryos/humans/societies are tomorrow's objects? We should and we do. It was mentioned in Chapter 3, the bio-philosphical themes that can be thrown out to a classroom – genes versus memes, the representations that are constitutive of humanity in books such as *So You Think You are Human?* by Fernandez-Armesto (2004) or Gray's (2002) *Straw Dogs*, or the STS of Haraway. The philosophical questions cannot be answered here, but writers such as Gray and Haraway bring animals into our political constituencies, and perhaps most notably ethicist Peter Singer broadens out the subjectivities to at least the animal world. Young people are said to have affinity with animal rights environment in education. But we do not listen enough to them, politically. In Singer, the field of subjective self shifts out to animals, but in shifting outwards, leaves behind newborn babies and the seriously mentally challenged. We tread very dangerously again.

But our anthropomorphic view of the world is most of what makes us human – how can it be any other way? Latour himself has called the human 'a weaver of morphisms' (Latour, 1993: 137), more humane than the *habitus*, the most central point between one axis of Nature and Subject/Society (where lies the domain of Discourse), and another of Essence and Existence (where lies Being). Of course this centrality of agency already sounds like Giddens' Third Way through the poles! The first plane has been more in focus here, but it is interesting that this somewhat solves the human question, taking in all future compositions of 'them' and 'us' as well as our potential strategies and appendages. This might not solve the conceptual issue for Latour however, of our humanity colonising the vastness of non-humanity out there. The conclusion has to be, yet again, that we see with human eyes, and with those only, the Kantian paradox which the newly emerging 'speculative realists' call 'correlationism' (Meillassoux, 2008). We can only find correlations between humans and non-humans without caring about their subjectivities.

If this is still an anthropogenic and anthropomorphic project, then good, we have kept the best parts of our humanity, and maybe all of it. Correlationsim it may be, but it may always be, as Bennet (2010: ix) calls it, the 'performative self-contradiction' – once we describe this new terrain of biology, we are humans still apparently calling the shots on post- and non-human forms of life, 'liberating' them. We perform what is for the best for us, for our social demonstration (Barry, 2001; Callon, 2004), for our future families, what is for *the good*; but we do so with one eye on the big picture or there is danger. As it is with environmental ethical matters, as it is with increased sexualisation of young people through imagery, as it is with neo-liberalist 'retail therapy'. Bennet may be trying to progress towards the ethics of things when she states: 'We need to cultivate a bit of anthropomorphism – the idea that human agency has some echoes in nonhuman nature – to counter the narcissism of humans in charge of the world' (Bennet, 2010: xvi). But we cannot escape Kant and human echoes they remain.

A critical classroom speaks to future parliament

The Irish Republic, currently as of 2013, is involved in a process of constitutional reform, moving towards the centenary year 2016, involving some form of citizens' assembly with elected parliamentarians and expertise, perhaps foreshadowing a complementary active citizenship as a contributory process of political representation. This is a defining period in Irish history now, where the Constitution needs to re-configure itself in its 'protection of the unborn', this elusive entity in Irish constitution, that is very Irish in its refracted meanings, a hybrid not mediated to the point of clarity in law.

Young people discussing potential natality of futures is one thing, apparently removed from Politics in the 'safety' of Education; but it should now be clearer that a couple deciding on ICSI procedures deals with two elements – it is both a naturally, liberal thing to do to allow them to choose a healthy child; but it is also looking back at history and knowing biopolitics is floating above their future decisions. Were the state to control this, history has taught us – and the UNESCO declaration from 1945 states clearly – we need to tread carefully. We must never return to the dark days of biological essentialism enshrined in state or transcendental utopias. Nor should regulation and legislation be in place that rigidly controls bodies. Ireland and other post-Catholic countries around the world are only too aware of the biopolitics of Catholic Church control, with state collusion, of the physical bodies of citizens, a state's children, but particularly its women, and how strategies that support life politics, that actually try to *bring back* the taboo, the messy, the embodied (in other words to name it, and make it happen), are dragged back by tactics of everyday practices (De Certeau, 1984) – in the Irish case, its laws and cultures[1]. However, relying completely on the reality of globalised neo-liberalism is not the answer either, as parents, individuals and future generations are placed at the mercy of markets. And so, it must be some form of Giddens' 'Third Way'. The tension of control will always be there. But above all it is practical manifestation of how life politics can work, from the earliest age a young citizen can engage as a potential parent, at latest just before voting age.

We have seen that the ideas coming out of STS, bioethics, and consideration of the non-humans and enhancement can provide this educational life politics. Lacking either of these misses an emerging historicity in contemporary theory. Lacking a platform for debate creates what Latour (2004a) constantly calls *speech impedimenta* – in this case meaning voices that are not capable of being offered as propositions from students or the structures, compositions and networks surrounding the whole traditional, slowly outmoded concept of biology.

More broadly for science governance, STS practitioners, scholars and science communication experts with STS leanings want a system that includes witnessing at various nodes in the networks. The processes of science and technology cannot be seen as linear anymore; technoscience drives science drives EC or NSF proposal text which drives technology which drives science which drives proposals. A now-famous diagram of high level upstream engagement between multiple actors shows a circular information flow, with expectation of co-production and change at each point, from agenda-setting of research, to concept formation, to R&D and dissemination (Jackson *et al.*, 2005). We have enough ICT, communications processes to allow this see-through science, as Wilsdon and Willis (2004) have called it, witnessing from proposal to desk to lab to inquiry meeting to town hall to classroom. The science in the classroom recorded during this book – maybe we cannot call it 'Science' – becomes part of witnessing and mediation from schools to government. But would arguments set up by Education and Outreach professionals where school kids see white coats working away in the background be any less of an argument around a concept than that in a classroom? Or could they get the opportunity to challenge the data that emerges from the screen? This is real citizen (bio)science. So while we have biologists Emma Teeling and Aoife McLysaght in Ireland, and Brian Cox in the UK, making public engagement a core part of what they do, a new breed, inspiring curiosity, it needs to go further than the curiosity of 'Did you know?'. The engagement needs to include the power of discourse and persuasion. Mystery and aesthetics also need to remain to preserve the rounded learning parts of us that need nuance, magic and kaleidoscope, for those of us who still want it. If *Caged Monkeys* and *Science Club* are what they say they are on the tin, then it is a smug club of caged primates! This is where Dawkins, Dennet and Winston are invaluable, being as they are contrarian parts of science communication and education. Remember, Attenborough and Dawkins contribute to the *process* of biology. And of course besides the aesthetics and process of biology, there are the practical real-life dilemmas. The example we looked at here, RGTs, was about deciding on what child, what form of life, what future, what control of choices of *the good*. How can education, traditionally fixed on knowledge of 'past certainties', realistically deal with these future-oriented complexities, of human and non-human entanglements, of future histories and moralities?

For parliaments and policy, this should not be a bureaucratic process. There would reasonably need to be good broadband, enough high speed ICT and good mediators on the ground, who are not just there for PR and the 'wow' factor of science (as institution or concept) but also there to provoke and to listen and to set a chain of acting inwards, impact from the outside. Always landing on a complex of Some Fractured Thing (Murphy, 2010), a hybrid non-Cartesian subject of Latour, the authentic self of Giddens. The Northern Irish Peace Process needed strong powers of persuasion and acceptance for a community and belief that went beyond 'mere' rhetoric. A certain kind of peace, if not felt by everyone, resulted, as structure of peaceful action. The structure is made by the actors themselves, enacted in the composition.

And so to present the logical conclusion to linking future opinion and politics to RGTs: we go back to the idea of 'stepping into discourse'. There are two ways that life politics can work in educational and political practice: 1) in the *event*, 'the value of the occasion', similar to Alfred North Whitehead's view of this concept (the performed, immanent, pedagogical interaction in class between teacher and students, or between politician and constituents); and 2) the Exchange-Gift idea, similar to Mauss – communication as reciprocity – give and take between family, society and generations. Mauss maintained that there is no pure form of gift without a contract.

A young 'agora' has a trusteeship, where lifestyles and future populations/ideas are explored; but present values are kept in evolving future constitutional set up. Future generations need a type of democratic trusteeship (Thompson, 2010). Can we represent future generations adequately? How can future generations speak to the present? They cannot speak directly. But there might be a type of dialogue. The concept of time, and particularly the future, has preoccupied sociologist Barbara Adam (1990) – how we humans perceive ourselves in the future, as old, with children, and for young people, with careers and adventures. Groves calls for a non-reciprocal responsibility, a future-oriented responsibility, which care for technological futures (Adam and Groves, 2007). We have a precedent for this in Ireland, called Dáil Na n'Og, 'Parliament of the Young', and environmental, bioethical and self-actualisation questions have been raised in debates in this organisation.

Mauss' unconditional non-exchange Gift, requires a contract, but calling it a social one might be misnomer, if we apply it at all times to a body-politic, one collection of subject-voters. A Maussian Gift, as we have touched on, draws on a linguistic tradition a *munus* that demands something in return (Campbell in Esposito, 2008: x). Entering into a community, in this case, regulating with present sacrifice for future millions or billions, is a heavy investment without visible return. If we protect the individual, there is immunisation, with risk of contagion, threatening these billions. Some form of visible-exchange constraint for the individual is necessary (Esposito, 2008). But it calls into question how and what to decide what is common. What is disease or undesirable (ibid.; Mol, 2002)? Who diagnoses to what end, on whom does it impact? What forms of life are created or, as Latour would ask, what connections are created that defines, then changes, the actants?

Politicians mediate on a surface level for the short term, *pro tempore*, because food, shelter and safety of the proximally known are paramount. This cannot be disputed. As Deirdre said: 'No-one knows until they're in that situation themselves' (Deirdre, ND 1, *The Gift* debates). We saw in Chapter 3 how Disch (2008) describes the performativity of our representatives, and we cannot demand much else (nor can we let politicians off the hook either). To prevent ultra-presentism, the future should be more involved in current democratic process and the reciprocal arrangement is that the education of the present creates a real democratic space for future populations and ideas to exist. Thompson doesn't think that a utilitarian argument would fare any better for future generations, even though such an ethical approach appears to care more for a block of persons rather than individuals, the many of unlimited world population growth against the few or the one (Thompson, 2010).

Let us examine a pragmatic solution. It is based on Rawls' idea of *just savings* – where present generations save for the rainy day of future populations. 'Do unto future generations as you would have past generations do unto you', he once said (Rawls, 1993: 274). We cannot presume to know what a future governance of people and things looks like. And by that logic, we cannot assume a moral structure contingent on a culture of today either. Here is where trusteeship comes in. The idea of trusteeship – like Locke's property/land principle, where 'present democratic sovereigns should leave their successors enough and as good democratic sovereignty as they themselves enjoy' (Thompson, 2010: 26), like a

'Tribune of the Plebs' in Roman times. The future trustee has a seat in the Parliament of Things that carries with it the essence and tradition of today, and will have the power to argue the rationale for these 'savings' (in reality, regulation) as an exchange-Gift to the then-present. She, the trustee, can be defeated if the future democratic or quasi-democratic houses so decide by their protocols. It must work that the present regulatory system sees it as future-protection from the past, not the competing values of that time (today), such as religion or neo-liberalism. In some ways, it is, in itself a kind of sequestration – for, and towards, the future. The future trustee is a defender of the Latourian democratic (or quasi-democratic) process, not the defender of future people, or non-humans for that matter. She will be a constant defender of the old laws, because 'these laws are there to protect you, not the then legislatures'. But she will also look to the further future, the 'F-representatives' (ibid. 30). The changing *zeitgeists* change the regulation of RGTs, but with knowledge of past and futures. The trustee will be a temporal witness.

This concept of the representative with just savings reduces presentism from current Parliament and also extends 'down' into educational reflection and argument. The trusteeship starts in the classroom, separated (in current thinking) from the political system. Students are 'twinned' with a future Parliament of Things. Therefore, what is elicited as opinion can also more effectively be seen as a natural by-product of action.

The idea is not so radical – a dialogue between the present and the future, from classroom to future governance. Today's students will be involved in a new politics, however formal/party political it will be. Future generations are silent, as are non-humans in this new political assembly. 'Society', as we construct it, is already involved. Political structures such as the Office of Science and Technology (POST) offer a glimpse of this, sub-politics (non-party) facilitating expert groups and special interest groups on emerging technologies. British (and Irish) citizens respond to each HfEA License Committee hearing via their website at the 'Have your say on conditions awaiting consideration' section. In the past Irish women needed travel to the UK for screening; but BEACON Fertility Clinic and NCMG now provide this service, albeit with different cost models. (Irish women seeking abortions still need to travel to the UK.) Any new political reform structure of the Upper House in Ireland, for example, might consider this long-term trusteeship of the political itself, a guarantee of representation of voices.

This is all about the relationship between opinion, individual action, collective action and representation. At one end of a spectrum you have tacit knowledge or practical wisdom. On the other, you have poise, which is a stepping into discourse, a formal discursive act, gathering oneself, giving opinion. Positioning in relation to frames can slide along this scale, but is closer to poise. The debater gathers herself for formal discourse; logic, reason, ethos, pathos all apply, but the non-verbal practices of local action, practical wisdom 'puts a spin' on the formal positioning. The implicit, unexplainable, embodied self gets in the way of reasoned discourse. How could it not?

To make non-human actors speak, they must be written into the play, with lines, as Latour does (Latour, 2010). Future humans then would also be considered in this assembly as not-yet-persons. For Giddens, a trusteeship means a subpolitics, a life politics, always there, formally beginning now. Schools could be linked to a youth parliament structure, which is a politics of an altogether different kind than now exists. A youth 'agora' has a trusteeship, where lifestyles and future populations/ideas are explored but present values are kept in the evolving future constitutional set up even if they appear in the future as archaic.

This work is a continuing response to what Bruno Latour asked for exactly twenty years ago. 'I have done my job as philosopher and constituent,' he said, 'By gathering together the scattered themes of a comparative anthropology. Others will be able to convene the Parliament of Things' (Latour, 1993: 145). There are many more steps to take to realise see-through science (Wilsdon and Willis, 2004), but even within education itself, this is just a tentative first step.

Science then is fundamentally rhetorical, communicative, argumentative, a creative subject, linked to others. Students need to come out of scientist mode at certain points and see how the biosciences are viewed elsewhere, from other perspectives. Let us mediate towards a simple 'elevator pitch' dilemma:

- Ask your students should we allow laissez-faire RGTs?
- Or pause, step back, debate, invoke technoscience in self-confrontation?

An interdisciplinary set of subjects and practices – 'transferrable skills' – are needed to complement its 'messiness'. New curricula should have such elements as Communication, Innovation, Creative Practice, Health & Wellbeing, Selling, Design, Laboratory Techniques, Abstract and Technique (texts). It might develop an understanding trusteeship of bioscience governance, that includes UNESCO, and crucially, education at all levels. From the abstract to the practical, from *technia* and ethics to logic, argument, rhetoric, *loquutio* – perhaps the presence of nearness itself, the technological sublime offers a way into 'truths', 'beauty', the good. Imagine a curriculum for fourteen to sixteen year olds that includes all this.

There is then the intriguing possibility of the sciences leading into alignments of contemporary politics, ethics, globalisation and innovation. The moralising nature concept comes to fruition. Science, one again, becomes Natural Philosophy, although 'Nature' has changed radically; in fact the old 'Nature' is radicalised.

The cast of characters of this book can be introduced in elements to the biology senior level classroom: our bio-orators in Chapter 2, but also the socialism of Bernal and Dobzhansky, the counter-biology of Lysenko inspiring political rejection of Darwinisim; the moral genetics of Gould and Lewontin; the biosociality of Uexküll and Esposito. Darwin is practically already there, because of his fame. Cox – with added televisual appeal – is too. Dawkins, the old protagonist that runs counter to the biosocial aims of this book, should be required reading (as should Jerry Ravetz's (2006) *The No-Nonsense Guide to Science*, as counterpoint).

The witness idea appears in the judicial system, but also in citizen juries in public engagement with science and technology, as it is in the gathering battlegrounds for the GM potato in Ireland (#gmpotato, 2013), the witness idea stretching back through Haraway (1997) and her 'Modest_Witness' to Shapin and Schaffer's 'virtual witnessing' (1985: 60). So what do we have for our classrooms to introduce life politics? We have:

- Non specialists – 'contact with the outside world' at every turn.
- Young speakers of slightly older age to the students.
- Facilitator-pedagogues, not to fear the blurring of the fictive, the legal, the personal, political and the scientific. This remains, in the absence of guidance and particularly exam change for teachers, the biggest challenge.
- Debate activities with marking schema for communication and science such as developed by the Wellcome Trust, Interdisciplinary Science and DCU and Debating Science Issues in Ireland as recorded here.
- TED videos of oration and rhetoric, as witnesses to anticipated future emerging technologies, with scenarios of the future, some in news report style, others more persuasive.
- A change in the culture of assessment of science.

Boyle was a performer, but he believed in the content of what he was saying; the structure affects the agency affects the virtual witnessing, affects structure, affects the agency *ad infinitum* (Shapin and Schaffer, 1985).

I will hope to have convinced the reader that there are large elements of science, and modern biology in particular, that can be taught more like a political ecology or a communications subject, full of argumentation and debate about ideas and practices, opening out to more witnessing and mediation of some expertise and none, some religious beliefs and none, and a good example of the usefulness of this approach can be demonstrated in the area of genetic enhancement of humans, likely to affect us all in future. This is not to say, and it is worth repeating, that biology and biotechnology are only discursive acts. Of course they are real, but reality is a composition – these acts are enacted in how we know them, and the value we bring to them (contra Dawkins, *et al.* who separate humanity from 'The Selfish Gene'). In embedding a political (small 'p'), communicative type of biology education, we can begin a system of guidance, control where needed and governance that links to wider political systems. Young people are not our governors – but on the other hand, this is not meant to be a weak system of politics either, as youth parliaments tend to be.

As with resource sustainability and climate change issues, we are saying to our 'frozen' future citizens (Adam, 1990), our children, this is our legacy, this is what we have left you. In all the flux of future change, there is a conservatism – there are things and ideas worth keeping and they are worth arguing for. Technology assessment requires education and science engagement to be meaningful, and the type of education that promotes critical thinking and future genetic scenarios will be the real impact for future citizens and future politics. With these types of exercises, it becomes, without a second thought, 'second nature' to think politically about nature, scientific enterprise and governance. Young people have increasingly more content, a *bricolage* of facts, opinion, horror and fun on which to construct identities of hope and enterprise. It makes sense that the biology subject is integrally linked to how these identities are chosen, or constructed. Biology – in its modern, pre-modern or multinaturalistic form (Latour, 2004a) – is *that* important.

Bíos need not bring with it *thanatos*. Biopolitics need not be thanatopolitics. Life politics is not death politics. We can appreciate the struggle for life and to live for each organism – and there may be much more than organic life – without falling into the trap of previous structures and regimes that used a will to power. Could we have the courage, with these tools, to bring forth a composition of, yes, *humanity*, in its continuing development and its realignment of older modes of existence, and to use them for a more inclusive common purpose? It is not ethics, but it is politics. It is *personal*, but still politics. And future generations will have many past actions at which to point and for which we will be accountable, so much decisive and unintentional action. Are we going to let *things* just happen?

Note

1 It has taken twenty years for a piece of legislation to be enacted in Ireland that allows abortion in extremely limited circumstances ('suicide ideation' in the mother). Before being passed into law, the 'Protection of Life in Pregnancy' Bill was debated through both houses of parliament (the Houses of the Oireachtais), then presented before the President who called on a special sitting of his advisory Council of State for advice. As any fertility expert of clinician in an emergency ward will testify, lives and life depend on much faster decision-making.

Appendix

Wellcome Trust film debate activities

The Gift Video Discussion: Questionnaire   Date ____

(Activity 3 from *The Gift* Teachers' Resource Pack, 1999, developed by Nigel Townsend, Y-Touring from Nicola Baldwin script)

School Code: ____   Male/Female (M/F):____

Character points of view

(1) JENNIFER is right to argue that even though she knows that there is a risk of her child inheriting Friedreich's ataxia, she wants 'nature' to take its course.

STRONGLY DISAGREE	DISAGREE	UNDECIDED	AGREE	STRONGLY AGREE
☐	☐	☐	☐	☐

CHANGE POSITION?

STRONGLY DISAGREE	*DISAGREE*	*UNDECIDED*	*AGREE*	*STRONGLY AGREE*
☐	☐	☐	☐	☐

(2) ANNIE is right, RYAN has a right to be tested because he has a right to know what's going to happen to him.

STRONGLY DISAGREE	DISAGREE	UNDECIDED	AGREE	STRONGLY AGREE
☐	☐	☐	☐	☐

CHANGE POSITION?

STRONGLY DISAGREE	*DISAGREE*	*UNDECIDED*	*AGREE*	*STRONGLY AGREE*
☐	☐	☐	☐	☐

(3) JENNIFER'S doctor is right to argue that Jennifer must do something now that she knows the risk.

STRONGLY DISAGREE	DISAGREE	UNDECIDED	AGREE	STRONGLY AGREE
☐	☐	☐	☐	☐

CHANGE POSITION?

STRONGLY DISAGREE	*DISAGREE*	*UNDECIDED*	*AGREE*	*STRONGLY AGREE*
☐	☐	☐	☐	☐

(4) RYAN'S doctor is right to argue that Ryan shouldn't be tested if there is no treatment available.

STRONGLY DISAGREE	DISAGREE	UNDECIDED	AGREE	STRONGLY AGREE
☐	☐	☐	☐	☐

CHANGE POSITION?

STRONGLY DISAGREE	*DISAGREE*	*UNDECIDED*	*AGREE*	*STRONGLY AGREE*
☐	☐	☐	☐	☐

(5) In selecting MARK, RYAN and JENNIFER were only doing what was best for him.

STRONGLY DISAGREE	DISAGREE	UNDECIDED	AGREE	STRONGLY AGREE
☐	☐	☐	☐	☐

CHANGE POSITION?

STRONGLY DISAGREE	*DISAGREE*	*UNDECIDED*	*AGREE*	*STRONGLY AGREE*
☐	☐	☐	☐	☐

(6) RYAN and JENNIFER were right to select MARK to be free of Friedreich's ataxia.

STRONGLY DISAGREE	DISAGREE	UNDECIDED	AGREE	STRONGLY AGREE
☐	☐	☐	☐	☐

CHANGE POSITION?

STRONGLY DISAGREE	*DISAGREE*	*UNDECIDED*	*AGREE*	*STRONGLY AGREE*
☐	☐	☐	☐	☐

(7) RYAN was right to select the sex of his child.

STRONGLY DISAGREE	DISAGREE	UNDECIDED	AGREE	STRONGLY AGREE
☐	☐	☐	☐	☐

CHANGE POSITION?

STRONGLY DISAGREE	*DISAGREE*	*UNDECIDED*	*AGREE*	*STRONGLY AGREE*
☐	☐	☐	☐	☐

(8) RYAN was right to select MARK to have the best qualities.

STRONGLY DISAGREE	DISAGREE	UNDECIDED	AGREE	STRONGLY AGREE
☐	☐	☐	☐	☐

CHANGE POSITION?

STRONGLY DISAGREE	*DISAGREE*	*UNDECIDED*	*AGREE*	*STRONGLY AGREE*
☐	☐	☐	☐	☐

Bibliography

#gmpotato (2013) Community of Inquiry. Online, available HTTP: http://www.gmpotato.com/wpcontent/uploads/2012/10/gmpotato_CommunityofInquiry_SummaryReport_Oct2012.pdf [accessed 13 September 2013].

Aagaard, K., Petrosino, J., Keitel, W., Watson, M., Katancik, J., Garcia, N., Patel, S., Cutting, M., Madden, T., Hamilton, H., Harris, E., Gevers, D., Simone, G., McInnes, P. and Versalovic, J. (2013) 'The Human Microbiome Project: strategy for comprehensive sampling of the human microbiome and why it matters'. *The FASEB Journal*, 27, 3: 1012–22.

Adam, B. (1990) *Time and Social Theory*. Cambridge, UK: Polity Press.

Adam, B. and Groves, C. (2007) *Future Matters: Action, Knowledge, Ethics*. Leiden, NL: Koninklijke Brill NV.

Adshead, M. and Millar. M. (2003) 'Ireland as Catholic Corporatist state: historical institutional analysis of healthcare in Ireland'. *Limerick Papers in Politics and Public Administration*, 5. Online, available HTTP: http://www.ul.ie/ppa/content/files/258567748.pdf [accessed 2 September 2013].

Agamben, G. (1998) *Homo Sacer: Sovereign Power and Bare Life*, trans. D. Heller-Roazen. Stanford, CA: Stanford University Press.

Agamben, G. (2000) *Means without End: Notes on Politics*, trans. V. Binetti and C. Casarino. Minneapolis, MN: University of Minneapolis Press.

Agamben, G. (2009) *What Is An Apparatus? And Other Essays*, trans. D. Kishick and S. Pedatella. Stanford, CA: Stanford University Press.

Aikenhead, G. (1996) 'Border crossings into the subculture of science'. *Studies in Science Education*, 27: 1–52.

Aikenhead, G. (2003) 'STS education: a rose by any other name', in R.Cross (ed.) *A Vision for Science Education: Responding to the Work of Peter J. Fensham*. New York: Routledge Press: 59–75.

Angrist, M. (2010) *Here Is a Human Being: At the Dawn of Personal Genomics*. New York: Harper.

Appleyard, B. (1999) *Brave New Worlds: Genetics and the Human Experience*. London: Harper Collins Publishers.

Appleyard, B. (2004) *Understanding the Present: an Alternative History of Science*. New York/London: Tauris Parke Paperbacks.

Arendt, H. (1958) *The Human Condition*. Chicago, IL: Chicago University Press.

Arendt, H. (2006) [1963] *Eichmann in Jerusalem: A Report on the Banality of Evil*. New York: Penguin Books.

Ask.fm (2013) (Homepage). Online, available HTTP: ask.fm [accessed 5 November 2013].

Atwood, M. (2003) *Oryx and Crake*. London: Virago.

Bakhtin, M. (1981) *The Dialogic Imagination: Four Essays*, M. M. Bakhtin. M. Holquist (ed.) Trans. C. Emerson and M. Holquist. Austin, TX: University of Texas Press.

Baldwin, N. (1999) *The Gift* script. London: Y-Touring Company. Online, available HTTP: http://www.geneticfutures.com/thegift/play/the_gift_script.pdf [accessed 8 September 2013].

Barberá, Ó., Zanón, B. and Pérez-Plá, J. F. (1999) 'Biology curriculum in twentieth-century Spain'. *Science Education* 83, 1: 97–111.

Barr, O. and Thompson, K. (2002) 'Trusting people with genetic information', in A. Gray, K. Lloyd, P. Devine, G. Robinson and D. Heenan (eds) *Social Attitudes in Northern Ireland: the 8th report*. London: Pluto Press: 70–86.

Barry, A. (2001) *Political Machines: Governing a Technological Society*. New York: The Athlone Press.

Barton, D. (2012) 'Diagnostic Molecular Genetics'. National Centre for Medical Genetics Lecture to University College Dublin Health Sciences First Years, 10 October.

Battles, J. (2002) 'Celtic gene link to female throat cancer'. *Sunday Times*, Home news, Scottish Edition, 8 September: 5.

Bauer, M. W. and Bonfadelli, H. (2002) 'Controversy, media coverage and public knowledge', in M. W. Bauer and G. Gaskell (eds) *Biotechnology: the Making of a Global Controversy*. London: Cambridge University Press, 149–75.

Bauer, M. W. and Gaskell, G. (2002a) 'Researching the public sphere of biotechnology', in M. W. Bauer and G. Gaskell (eds) *Biotechnology: the Making of a Global Controversy*. London: Cambridge University Press, 1–17.

Bauer, M. W. and Gaskell, G. (2002b) 'The biotechnology movement', in M. W. Bauer and G. Gaskell (eds) *Biotechnology: the Making of a Global Controversy*. London: Cambridge University Press, 379–404.

Beauchamp, T. L. and Childress, J. F. (1994) *Principles of Biomedical Ethics*. New York: Oxford University Press.

Beck, U. (1992) *Risk Society: Towards a New Modernity*. London: Sage.

Beck, U. (1994) 'The reinvention of politics', in U. Beck, A. Giddens and S. Lash (eds) *Reflexive Modernisation: Politics, Tradition and Aesthetics in the Modern Social Order*. Cambridge, UK: Polity Press, 1–55.

Bennet, J. (2010) *Vibrant Matter: a Political Ecology of Things*. Durham, NC: Duke University Press.

Benton, T. (1991) 'Biology and social science: why the return of the repressed should be given a (cautious) welcome'. *Sociology*, 25, 1: 1–29.

Bergson, H. (1914) *Gifford Lectures, 1914: Problem of personality. Lecture* 2. Edinburgh: University of Edinburgh Press.

Billig, M. (1987) *Arguing and Thinking: a Rhetorical Approach to Social Psychology*. Cambridge, UK: Cambridge University Press.

Bloor, D. (1991) *Knowledge and Social Imagery* (2nd ed.). Chicago, IL: Chicago University Press.

Blue, L. (2007) 'The mortification of James Watson' *Time* October. Online, available HTTP: http://content.time.com/time/health/article/0,8599,1673952,00.html [accessed 9 August 2013].

Boisvert, R. D. (2010) Convivialism: a philosophical manifesto'. *The Pluralist*, 5, 2: 57–68.

Boltanski, L. (1999) *Distant Suffering: Morality, Media, and Politics*. Cambridge, UK: Cambridge University Press.

Boltanski, L. and Chiapello, E. (2006) *The New Spirit of Capitalism* (2nd ed.). London: Verso.

Boltanski, L. and Thévenot, L. (2006) *On Justification: Economies of Worth*, trans. C. Porter. New Jersey: Princeton University Press.

Bourdieu, P. (1990) *The Logic of Practice*. Cambridge, UK: Polity Press.

Bourdieu, P. and Passeron, J.C. (1977) *Reproduction in Education, Society, and Culture*, trans. R. Nice. London: Sage Publications.

Bourdieu, P. and Wacquant, L. (1992) *An Invitation to Reflexive Sociology*. Cambridge, UK: Polity Press.

Broadcasters' Audience Research Board (2012) BARB Bulletin. Online, available HTTP: http://www.barb.co.uk [accessed 10 June 2013].

Brossard, D. and Scheufele, D. A. (2013) 'Science, new media and the public'. *Science*, 339, 6115: 40–1.

Brummett, B. (2013) 'What popular films teach us about values: locked inside with the rage virus'. *Journal of Popular Film and Television*, 41, 2: 61–7.

Buchanan, A., Brock, W. B., Daniels, N. and Wikler, D. (2000) *From Chance to Choice: Genetics and Justice*. Cambridge, UK: Cambridge University Press.

Bucchi, M. (2004) *Science in Society: an Introduction to Social Studies of Science*. London and New York: Routledge.

Buckingham, D. (2003) 'Children and television: a critical overview of the research', in V. Nightingale and K. Ross (eds) *Critical Readings: Media and Audiences*. Maidenhead: Open University Press, 163–80.

Bultitude, K. (2011) 'The why and the how of science communication', in P. Rosulek (ed.) *Science Communication*. Pilsen, CZ: European Commission.

Bunreacht na hÉireann (2012) Constitution of Ireland. Online, available HTTP: http://www.taoiseach.gov.ie/eng/Historical_Information/The_Constitution/ [accessed 9 September 2013].

Buraga, S-C. (2013) Society. Online e-painting, available HTTP: http://profs.info.uaic.ro/~busaco/paint/ [accessed 20 October 2013].

Butler, J. (1990) *Gender Trouble: Feminism and the Subversion of Identity*.London and New York: Routledge.

Butler, J. (1996) 'Performativity's social magic', in T. Schatzki and W. Natter (eds) *The Social and Political Body*. New York/London: The Guildford Press, 29–47.

Callon, M. (2004) 'Europe wrestling with technology'. *Economy & Society*, 31, 1: 121–34.

Callon, M. and Latour, B. (1981) 'Unscrewing the Big Leviathan: how actors macrostructure reality and how sociologists help them to do so,' in K. D. Knorr-Cetina and A. V. Cicourel (eds) *Advances in Social Theory and Methodology: Toward an Integration of Micro- and Macro-Sociologies*. Boston, MA: Routledge and Kegan Paul, 277–303.

Calvino, I. (1987) 'Cybernetics and Ghosts', in I. Calvino (ed.) *The Literature Machine* (2nd ed.). London: Vintage Classics.

Campbell, T. C. (2011) *Improper Life: Technology and Biopolitics from Heidegger to Agamben*. Minneapolis, MN and London: University of Minnesota Press.

Canguilhem, G. (1979) [1945] *Le Normal et le pathologique*. Paris: PUE.

Canguilhem, G. (1992) 'Machine and Organism', trans. Mark Cohen and Randall Cherry, in J. Crary and S. Kwinter (eds) *Incorporations*. New York: Zone Books.

Casero-Ripollés, A. (2012) 'Beyond newspapers: news consumption among young people in the digital era'. *Comunicar*, 20, 39: 151.

Cassata, F. (2011) *Building the New Man: Eugenics, Racial Science and Genetics in Twentieth Century Italy*, trans E. O'Loughlin. Budapest: Central European University Press.

Cassidy, L. (2006) 'Court says frozen embryos "not unborn"'. *Irish Times*. Online, available HTTP: http://www.ireland.com/newspaper/breaking/2006/1115/breaking36.htm [accessed 13 March 2007].

Cavarero, A. (2005) *For More than One Voice: Toward a Philosophy of Vocal Expression*, trans. P. A. Kottman. Stanford, CA: Stanford University Press.

Centre for Bioethics and Human Dignity (2013) G12 Country Regulations of Assisted Reproductive Technologies, Online, available HTTP: http://cbhd.org/content/g12-country-regulations-assisted-reproductive-technologies [accessed 15 August 2013].

Chaffee, S. H. and Yang, S-M. (1990) 'Communication and political socialisation', in O. Ichilov (ed.) *Political Socialisation, Citizenship Education, and Democracy*. New York: Teachers College Press.

Challand, C. (2003) 'We want a British designer baby to save our sick son'. *Mail on Sunday*, 22 June: 32.

Chang, L. J., Chen, S. U., Tsai, Y. Y., Hung, C. C., Fang, M. Y., Su, Y.N. and Yang, Y.S. (2011) 'An update of preimplantation genetic diagnosis in gene diseases, chromosomal translocation, and aneuploidy screening'. *Clinical and Experimental Reproductive Medicine*, 38, 3: 126–34.

Chowning, J. T., Griswold, J. C., Kovarik, D. N. and Collins, L. (2012) 'Fostering critical thinking, reasoning, and argumentation skills through Bioethics Education'. *PLoS One*, 7, 5: e367917.

Cobern, W. (1996) 'Worldview theory and conceptual change in science education'. *Science Education*, 80, 5: 579–610.

Collins, H. M. and Evans, R. (2002) 'The third wave of Science Studies: Studies of expertise and experience'. *Social Studies of Science*, 32, 2: 235–96.

Commission for Assisted Human Reproduction (CAHR) (2005) *Report of the Commission on Assisted Human Reproduction*. Online, available HTTP: http://www.dohc.ie/publications/pdf/cahr.pdf?direct=1. [accessed 8 July 2013].

Costa, V. B. (1995) 'When science is "another world": relationships between worlds of family, friends, school, and science'. *Science Education*, 79, 3: 313–33.

Couldry, N. (2003) *Media Rituals: a Critical Approach*. London and New York: Routledge.

C-Spanvideo.org (1992) Online, available HTTP: http://www.c-spanvideo.org/program/185880-1 [accessed 5 September 2013].

Curriculumonline.ie (2003) Leaving Certificate Biology Syllabus (Higher Level and Oridinary Level). Online, available HTTP: http://www.curriculumonline.ie/uploadedfiles/pdf/lc_biology_sy.pdf. [accessed 29 October 2013].

Davies, B. and Harre, R. (1997) 'Positioning: the discursive production of selves'. Online, available HTTP: http://www.massey.ac.nz/~alock/position/position.htm. [accessed 2 November 2013].

de Certeau, M. (1984) *The Practice of Everyday Life*, trans. S. F. Randall. Berkeley and Los Angeles, CA: University of California Press.

de Laat, B. (2000) 'Scripts for the future: using innovation studies to design foresight tools', in N. Brown, B. Rappert Brian and A. Webster (eds) *Contested Future: A Sociology of Prospective Techno-science*. Aldershot: Ashgate, 175–208.

Deleuze, G. and Guattari, F. (1987) *A Thousand Plateaus*, trans. B. Massumi. Minneapolis, MN: University of Minnesota Press.

den Boer, D., Rip, A. and Speller, S. (2009) 'Scripting possible futures of nanotechnologies: a methodology that enhances re?exivity'. *Technology in Society*, 31: 295–304.

Devins, N. (2009) 'How Planned Parenthood v. Casey (pretty much) settled the Abortion Wars'. *Yale Law Journal*, 118: 1318–54.

Diamond, Commissioner of Patents and Trademarks v. Chakrabarty (1980) Online. 447 U.S. Supreme Court 303, 206 USPQ 193 Online, available HTTP: http://digital-law-online.info/cases/206PQ193.htm [accessed 12 August 2013].

Disch, L. (2008) 'Representation as "spokespersonship": Bruno Latour's political theory'. *Parallax*, 14: 3, 88–100.

Disch, L. (2010) '"Faitiche"-izing the people: what representative democracy might learn from science studies', in B. Braun and S. J. Whatmore (eds) *Political Matter: Technoscience, Democracy and Public Life*. Minneapolis, MN and London: University of Minnesota Press, 267–96.

Dolgin, J. (1990) 'Status and contract in feminist legal theory of the family: a reply to Bartlett'. *Women's Rights Law Reporter*, 12: 103–13.

Dorkin, R. (2000) 'Playing God: genes, clones and luck', in *Sovereign Virtue: The Theory and Practice of Equality*. Cambridge, MA: Harvard University Press.

Downes, J. (2005) 'Mixed reaction to "abortion" question'. *Irish Times*, Exam Times, 15 June: 6.

Driver, R. Leach, J., Millar, R. and Scott, P. (1996) *Young People's Images of Science*. London: Open University Press.

Driver, R., Newton, P. and Osborne, J. (2000) 'Establishing the norms of scientific argumentation in classrooms'. *Science Education*, 84, 3: 287–312.

Duddy, T. (2002) *A History of Irish Thought*. London: Routledge.

Durodié, B. (2003) 'Limitations of public dialogue in science and the rise of new "experts"'. *Critical Review of International Social and Political Philosophy*, 6, 4: 82–92.

Eder, K. (1996) *The Social Construction of Nature: a Sociology of Ecological Enlightenment*. London: Sage Publications.

Edwards, J. (1999a) 'Explicit connections: ethnographic enquiry in north-west England', in J. Edwards, S. Franklin, E. Hirsch, F. Price and M. Strathern (eds) *Technologies of Procreation: Kinship in the Age of Assisted Reproduction* (2nd ed.). London and New York: Routledge, 60–85.

Edwards, J. (1999b) 'Clones – who are they?' in J. Edwards, S. Franklin, E. Hirsch, F. Price and M. Strathern (eds) *Technologies of Procreation: Kinship in the Age of Assisted Reproduction* (2nd ed.). London and New York: Routledge, 86–90.

Edwards, J. (2002) 'Taking "public understanding" seriously'. *New Genetics and Society*, 21, 3: 315–25.

Einsiedel, E. (2005) 'Editorial: Of Publics and Science'. *Public Understanding of Science*, 16, 1: 5–6.

Embryo Protection Act (1990) Online. Gesetz zum Schutz von Embryonen (Embryonenschutzgesetz – ESchG), Embryonenschutzgesetz (ESchG). Online, available HTTP: http://www.auswaertiges-amt.de/cae/servlet/contentblob/480804/publicationFile/5162/EmbryoProtectionAct.pdf [accessed 28 July 2013].

Entman, R. M. (1993) 'Framing: toward clarification of a fractured paradigm', *Journal of Communication,* 43, 4: 51–8.

Erickson, M. (2005) *Science, Culture and Society: Understanding Science in the 21st Century.* Cambridge, UK: Polity Press.

Esposito, R. (2002) *Immunitas:Protezione e Negazione della Vita.* Turin: Einaudi.

Esposito, R. (2008) *Bios: Biopolitics and Philosophy,* trans. T. Campbell. Minneapolis, MN and London: University of Minnesota Press.

European Commission (1997) *Europeans and Modern Opinions about Biotechnology: Eurobarometer 46.1.* Brussels: European Commission.

European Commission Eurostat (2012) Online, available HTTP: http://epp.eurostat.ec.europa.eu/statistics_explained/index.php/R_%26_D_expenditure [accessed 4 September 2013].

Evans, G. and Durant, J. (1995) 'The relationship between knowledge and attitudes in the public understanding of science in Britain'. *Public Understanding of Science* 4, 1: 57–74.

Expert Group on Future Skills Needs (2013) *Monitoring Ireland's Skills Supply: Trends in Education and Training Outputs.* Dublin: Forfas.

Fairclough, N. (2003) *Analysing Discourse: Textual Analysis for Social Research.* London and New York: Routledge.

Fernandez-Armesto, F. (2004) *So You Think You're Human?: A Brief History of Humankind.* Oxford: Oxford University Press.

Ferree, M. M., Gamson, W. A., Gerhards, J. and Rucht, D. (2002) *Shaping Abortion Discourse: Democracy and the Public Sphere in Germany and the United States.* Cambridge, UK: Cambridge University Press.

Fleck, L. (1979) [1935] *Genesis and Development of Scientific Fact,* trans. F. Bradley and T.J. Trenn. Chicago, IL and London: University of Chicago Press.

Foucault, M. (1978) *The History of Sexuality, vol. 1: The Will to Knowledge.* London: Penguin.

Foucault, M. (1995) [1977]) *Discipline and Punish: the Birth of the Prison,* trans. A. Sheridan. New York: Vintage Books.

Franklin, S. (1999) 'Postmodern procreation: a cultural account of assisted reproduction', in J. Edwards, S. Franklin, E. Hirsch, F. Price and M. Strathern (eds) *Technologies of Procreation: Kinship in the Age of Assisted Reproduction* (2nd ed.) London and New York: Routledge, 323–45.

Franklin, S. (2000) 'Life itself: Global nature and the genetic imaginary', in S. Franklin, C. Lury and J. Stacey (eds) *Global Nature, Global Culture.* London: Sage, 188–227.

Franklin, S. (2003) Rethinking nature-culture: anthropology and the new genetics. *Anthropological Theory,* 3, 1: 65–85.

Franklin, S. (2006) 'The cyborg embryo: our path to transbiology'. *Theory Culture Societ,* 23, 7–8: 167–87.

Freire, P. (1989)[1970] *The Pedagogy of the Oppressed,* trans. M.B. Ramos. New York: Continuum Publishing Company.

Freeman, S. (2011) *Biological Science* (4th ed.). San Francisco, CA: Pearson.

Fukuyama, F. (2002) *Our Posthuman Future.* London: Profile Books.

Fuller, S. (1997) *Science.* Buckingham: Open University Press.

Fuller, S. (2007) *New Frontiers in Science and Technology Studies.* Cambridge, UK: Polity Press.

Fuller, S. (2010) *Science: The Art of Living.* Durham, UK: Acumen.

Fuller, S. (2011) *Humanity 2.0: What it Means to be Human Past, Present and Future.* New York: Palgrave Macmillan.

Fuller, S. and Collier, J. H. (2008) *Philosophy, Rhetoric and the End of Knowledge: a New Beginning for Science and Technology Studies.* New Jersey: Lawrence Erlbaum and Associates.

Gamson, W. A. (1992) *Talking Politics.* Cambridge, UK/New York: Cambridge University Press.

Gamson, W. A. and Modigliani, A. (1989) 'Media discourse and public opinion on nuclear power: a constructivist approach'. *The American Journal of Sociology*, 95, 1: 1–37.

Gaskell, G. (2003) *Ambivalent GM Nation: Public Attitudes to Biotechnology in the UK, 1991–2002: Life Sciences in European Society Report.* London: London School of Economics.

Gaskell, G., Thompson, P. and Allum, N. (2002) 'Worlds apart?: public opinion in Europe and the USA', in M. W. Bauer and G. Gaskell (eds) *Biotechnology – the Making of a Global Controversy.* London: Cambridge University Press, 351–75.

Gaskell, G., Allum, N., Bauer, M., Durant, J., Allansdottir, A., Bonfadelli, H., Boy, D., De Cheveigné, S., Fjaestad, B., Gutteling, J. M., Hampel, J., Jelsøe, E., Jesuino, J. C., Kohring, M., Kronberger, N., Midden, C., Nielsen, T.H., Przestalski, A., Rusanen, T., Sakellaris, G., Torgersen, H., Twardowski, T. and Wagner W. (2000) 'Biotechnology and the European Public'. *Nature Biotechnology*, 18, 9: 935–38.

Gattaca (1997) Movie script. Online, available HTTP: http://www.dailyscript.com/scripts/gattaca_early.html [accessed 4 July 2013].

Gibbons, M. (1999) 'Science's new social contract with society', *Nature*, 402, C81.

Gibbons, M., Limoges, C., Nowotny, H., Schwartzman, S., Scott, P. and Trow, M. (1994) *The New Production of Knowledge: the Dynamics of Science and Research in Contemporary Society.* London: Sage.

Giddens, A. (1991) *Modernity and Self-identity: Self and Society in the Late Modern Age.* Cambridge, UK: Polity Press.

Giddens, A. (1994) 'Living in a post-traditional society', in U. Beck, A. Giddens and S. Lash (eds) *Reflexive Modernisation: Politics, Tradition and Aesthetics in the Modern Social Order.* Cambridge, UK: Polity Press, 56–109.

Giddens, A. (2009) *The Politics of Climate Change.* Cambridge, UK: Polity Press.

Giddens, A. (2012)[1990] *The Consequences of Modernity.* Cambridge, UK: Polity Press

Gieryn T. F. (1983) 'Boundary-work and the demarcation of science from non-science: strains and interests inprofessional ideologies of scientists'. *American Sociological Review*, 48, 6: 781–95.

Gift, The (1999) [Video: VHS]. London, Y-Touring Company.

Giroux, H. A. (1992) *Border Crossings: Cultural Workers and the Politics of Education.* London and New York: Routledge.

Giroux, H. A. (1994) 'Doing cultural studies: youth and the challenge of pedagogy'. *Harvard Educational Review*, 64, 3: 278–308. Online, available HTTP: http://www.henryagiroux.com/online_articles/doing_cultural.htm [accessed 6 November 2013].

Giroux, H. (1997a) *Education and Cultural Studies: Toward a Performative Practice.* London and New York: Routledge.

Giroux, H. (1997b) *Pedagogy and the Politics of Hope.* Boulder, CO:Westview Press.

Gitlin, T. (1980) *The Whole World is Watching: Mass Media in the Making and Unmaking of the New Left.* Los Angeles and London: University of California Press.

Goffman, E. (1974) *Frame Analysis: an Essay on the Organization of Experience*. New York: Harper Torchbooks.

Goffman, E. (1990) [1959] *The Presentation of Self in Everyday Life*. London: Penguin Books.

Goodfield, J. (1977) *Playing God: Genetic Engineering and the Manipulation of Life*. New York: Random House.

Gottweiss, H. (1998) *Governing Molecules: The Discursive Politics of Genetic Engineering in Europe and United States*. Cambridge, MA: MIT Press.

Gould, S. J. (1996) [1981] *The Mismeasure of Man*. New York: W. W. Norton & Company.

Gray, J. (2002) *Straw Dogs: Thoughts on Humans and Other Animals*. London: Granta Books.

Groves, C. (2006) 'Technological Futures and Non-Reciprocal Responsibility'. Paper presented to the Fourth International Conference on New Directions in the Humanities, University of Carthage, Tunis, Tunisia, 3–6 July.

Guston, D. H. and Sarewitz, D. (2002) 'Real-time technology assessment'. *Technology in Society*, 24: 93–109.

Habermas, J. (2003) *The Future of Human Nature*, trans. H. Beister, M. Pensky and W. Rehg. Cambridge, UK: Polity Press.

Hall, S. (1996) 'Introduction: who needs cultural identity?' in S. Hall and P. du Gay (eds) *Questions of Cultural Identity*. Thousand Oaks, CA and London: Sage, 1–17.

Hansen, A. (1994) 'Journalistic practices and science reporting in the British press'. *Public Understanding of Science*, 3, 2, 111–34.

Haraway, D. (1976) *Crystals, Fabrics and Fields: Metaphors of Organicism in Twentieth-century Developmental Biology*. New Haven, CT: Yale University Press.

Haraway, D. (1991) 'A cyborg manifesto: science, technology, and socialist-feminism in the late twentieth century', in D. Haraway (ed.) *Simians, Cyborgs and Women: the Reinvention of Nature*. New York: Routledge, 149–81.

Haraway, D. (1997) *Modest_Witness@Second_Millennium.FemaleMan©_Meets_ OncoMouse*. London and New York: Taylor & Francis Group.

Harding, S. (1986) *The Science Question in Feminism*. Ithaca, NY: Cornell University Press.

Harm, M. and Wiehle, H. (1942) *Lebenskunde für Mittelschulen. Fünfter Teil. Klasse 5 für Mädchen*. Online, trans. R. Bytwerk Halle: Hermann Schroedel Verlag, available HTTP: http://www.calvin.edu/academic/cas/gpa/textbk01.htm [accessed 5 September 2013].

Harre, R. and van Langenhove, L. (1999) 'The dynamics of social episodes', in R. Harre and L. van Langenhove (eds) *Positioning Theory: Moral Contexts of Intentional Action*. Oxford/Massachusetts: Blackwell Publishers Ltd, 1–13.

Harrington, A. (1996) *Reenchanted Science: Holism in German Culture from Wilhelm II to Hitler*. New Jersey: Princeton University Press.

Hayry, M. (2003) 'Philosophical arguments for and against human reproductive cloning'. *Bioethics*, 17(5-6): 447–59.

Heidegger, M. (1977) 'Letter on Humanism', in D. Farrell Krell (ed.) *Basic Writings from Being and Time (1927) to the Task of Thinking (1964)*. San Fransciso: Harper San Fransciso, 193–242.

Heidegger, M. (1992)[1942] *Parmenides*, trans. A. Schuwer and R. Rocjewicz. Bloomington, IN: Indiana University Press.

Heidegger, M. (2001) *Zollikon Seminars: Protocols, Conversations, Letters*, (ed.) M. Boss. Evanston, IL: Northwestern University Press.

Hilgartner, S. (1990) 'The dominant view of popularization: conceptual problems, political issues'. *Social Studies of Science*, 20, 3: 519–39.

Hilgartner, S. (2000) *Science on Stage: Expert Advice as Public Drama*. Stanford, CA: Stanford University Press.

Hill, E. W., Jobling, M. A. and Bradley, D. G. (2000) 'Y-chromosome variation and Irish origins'. *Nature*, 404, 23: 351–52.

Hipkins, R. (2002) *Commonsense, Trust, and Science: How Patterns of Beliefs and Attitudes to Science Pose Challenges for Effective Communication*. Wellington, New Zealand: Ministry of Research, Science and Technology.

Hoban, T.J., Woodrum, E. and Czaja, R. (1992) 'Public opposition to genetic engineering'. *Rural Sociology*, 5, 4: 476–93.

Hobbes, T. (1651) *Leviathan, or the Matter, Forme, & Power of a Common-wealth Ecclesiasticall and Civill*. [Online]. McMaster University Archive of the History of Economic Thought. London: Andrew Crooke, available HTTP: http://socserv.mcmaster.ca/econ/ugcm/3ll3/hobbes/Leviathan.pdf [accessed 1 September 2013].

Hornig Priest, S. (2001) *A Grain of Truth: the Media, The Public and Biotechnology*. Lanham, Boulder, New York, Oxford: Rowman and Littlefield Publishers Inc.

Hornig Priest, S. and Ten Eyck. T. (2003) 'News coverage of biotechnology debates'. *Society* 40, 6: 29–35.

Horst, M. (2008) 'The laboratory of public debate: understanding the acceptability of stem cell research'. *Science and Public Policy*, 35, 3: 197–205.

Human Fertilisation and Embryology Authority (HFEA) Ethics and Law Committee (2003) *CORE v HFEA: Implications for PGD Licensing*. Online, available HTTP: http://www.hfea.gov.uk/docs/ELC_4_july03.pdf [accessed 29 July 2013].

Hunter, G.W. (1914) *A Civic Biology: Presented in Problems*. New York: American Book Company.

Huxley, A. (1994) [1932] *Brave New World*, London: Flamingo.

Huxley, L. (2012) [1903] *Life and Letters of Thomas Henry Huxley Vol 1*. Cambridge, US: Cambridge University Press.

Huxley, T. H. and Martin, N. H. (1875) *Course of Practical Instruction in Elementary Biology*. London: Macmillan.

If...Cloning Could Cure Us (2004a) BBC 2, 16 December.

If...Cloning Could Cure Us (2004b) Online, BBC website, available HTTP: http://news.bbc.co.uk/2/hi/programmes/if/4065719.stm [accessed 30 July 2013].

Inhorn, M. C. and Birenbaum-Carmeli, D. (2008) 'Assisted Reproductive Technologies and culture change'. *Annual Review of Anthropology*, 37: 177–96.

Irwin, A. (2006) 'The politics of talk: coming to terms with the "new" scientific governance'. *Social Studies of Science*, 36, 2: 299–320.

Island of Dr. Moreau, The (1996) [Film]. Directed by John Frankenheimer, USA: New Line Cinema.

Jackson, R., Barbagallo, F. And Haste, H. (2005) 'Strengths of public dialogue on science-related issues'. *Critical Review of International Social and Political Philosophy*, 8, 3: 349–58.

James, W. (1996)[1907] *Essays in Radical Empiricism*. London: University of Nebraska Press.

Jarman, R. and McClune, B. (2002) 'A survey of the use of newspapers in science instruction by secondary teachers in Northern Ireland'. *International Journal of Science Education*, 24, 10: 997–1020.

Jasanoff, S. (2005) *Designs on Nature: Science and Democracy in Europe and the United States*. New York: Princeton University Press.

Jensen, E. (2009) 'Genetic futures and the media'. Online, eLS- Encyclopedia of Life Sciences, available HTTP: http://www.els.net/WileyCDA/ElsArticle/refId-a0005863.html. [accessed 7 November 2013].

Jiménez-Aleixandre, M. P., Bugallo Rodríguez, A. and Duschl, R. A. (2000) 'Doing the lesson or doing science: argument in high school genetics'. *Science Education*, 84, 6: 757–92.

Jones, S. (1996) *In the Blood: God, Genes and Destiny*. London: Flamingo.

Kass, L. R. (1985) *Toward a More Natural Science: Biology and Human Affairs*. New York: Macmillan.

Kazan, T. S. (2005) 'Dancing bodies in the classroom: moving toward an embodied pedagogy'. *Pedagogy: Critical Approaches to Teaching Literature, Language, Composition, and Culture*, 5, 3: 379–408.

Keller, E. F. (2002) *Making Sense of Life: Explaining Biological Development with Models, Metaphors, and Machines*. Cambridge, MA: Harvard University Press.

Kelly, O. (2011) 'Deconstructing "Grown versus Made:" a Derridean perspective on cloning' [Online]. *Journal of Philosophy: A Cross Disciplinary Inquiry*, 7, 16: 42–52. Online, available HTTP: http://www.questia.com/library/1G1-270365636/deconstructing-grown-versus-made-a-derridean-perspective [accessed 6 November 2013].

Kenny, M. (2002) 'Tears are a bad basis for manufacturing babies', *The Telegraph*, 6 August. Online, available HTTP: http://www.telegraph.co.uk/comment/personal-view/3580074/Tears-are-a-bad-basis-for-manufacturing-babies.html [accessed 30 July 2013].

Kfoury C. (2007) 'Therapeutic cloning: promises and issues'. *MJM: McGill Journal of Medicine*, 10(2):112–20.

Kitchin, R. and Kneale, J. (2001) 'Science fiction or future fact?: exploring imaginative geographies of the new millennium'. *Progress in Human Geography*, 25, 1: 19–35.

KK Women's and Children's Hospital (2013) (Homepage) Online, available HTTP: http://www.kkh.com.sg/ [accessed 6 September 2013].

Klein, K. (2010) '"New Authoritarianism" in China: Political Reform in the One Party State'. *TELOS*, 151: 30–56.

Kohlberg, L. (1981) *Essays on Moral Development, Vol. I: the philosophy of moral development*. New York: Harper and Row.

Kohring, M. and Matthes, J. (2002) 'The face(t)s of biotech in the nineties: how the German press framed modern biotechnology'. *Public Understanding of Science*, 11: 143–54.

Kolstø, S.D. (2001) 'To trust or not to trust …': pupils' ways of judging information encountered in a socio-scientific issues'. *International Journal of Science Education*, 23: 877–901.

Krishna, V. V. (1992) 'The Colonial "Model" and the Emergence of National Science in India: 1876-1920', in P. Petitjean *et al.* (eds) *Science and Empires: Boston Studies in the Philosophy of Science*, 136: 57–72.

Kuhlmann, A. (2001) *Politik des Lebens, Politik des Sterbens*. Berlin: Alexander Fest Verlag.

Laclau E. and Mouffe, C. (2001) *Hegemony and Socialist Strategy: Towards a Radical Democratic Politics*. London and New York: Verso.

Lash, S. (1994) 'Reflexivity and its doubles', in U. Beck, A. Giddens and S. Lash (eds) *Reflexive Modernisation: Politics, Tradition and Aesthetics in the Modern Social Order*, Cambridge, UK: Polity Press, 110–73.

Latour, B. (1988) *The Pasteurization of France.* Cambridge, MA: Harvard University Press.

Latour, B. (1993) *We Have Never Been Modern*, trans. C. Porter. Cambridge, MAS: Harvard University Press.

Latour, B. (2003) 'What if we *talked* politics a little?' *Contemporary Political Theory*, 2: 143–64, trans. L. Libbrecht. First published in French in *Politix* 15, 58: 143–66 (2002).

Latour, B. (2004a) *The Politics of Nature: How to Bring the Sciences into Democracy*, trans. C. Porter. Cambridge, MA: Cambridge University Press.

Latour, B. (2004b) 'How to talk about the body?: the normative dimension of science'. *Body & Society*, 10, 2–3: 205–29.

Latour, B. (2005a) 'From realpolitik to dingpolitik or how to make things public', in B. Latour and P. Weibel (eds) *Making Things Public: Atmospheres of Democracy.* Cambridge, MA: MIT Press, 14–43.

Latour, B. (2005b) *Reassembling the Social.* Oxford: Oxford University Press.

Latour, B. (2010) 'An attempt at a "Compositionist Manifesto"'. *New Literary History,* 41: 471–90.

Latour, B. (2012) 'From critique to composition'. Workshop to Celsius Research Group, Dublin City University, 17 February. Online, available HTTP: http://www.youtube.com/watch?v=-02aCvQ-HFs [accessed 29 July 2013].

Latour, B. (2013) *An Inquiry into Modes of Existence: An Anthropology of the Moderns.* Cambridge, MA: Harvard University Press.

Latour, B., Eliasson, O., Elie During, E. and Canales, J. (2010) Le débat Bergson/Einstein, Conference at Centre Pompidou, Paris, 21 June. Online, available HTTP: http://www.centrepompidou.fr/cpv/ressource.action?param.id=FR_R-d715393a5581f1cb4687486429b62a4d&param.idSource=FR_E-d715393a5581f1cb4687486429b62a4d [accessed 2 November 2013].

Latour, B., Jensen, P., Venturini, T., Grauwin, S. and Boullier, D. (2012) '"The whole is always smaller than its parts" – a digital test of Gabriel Tardes' monads'. *The British Journal of Sociology*, 63, 4: 590–615.

Latour, B. and Woolgar, S. (1979) *Laboratory Life: The Social Construction of Scientific Facts.* London: Sage.

Laveaga, G. S. (2009) *Jungle Laboratories: Mexican Peasants, National Projects, and the Making of the Pill.* Durham, NC: Duke University Press.

Leach, J., Lewis, J., Drover, R. and Wood-Robinson, C. (1996) Working Paper 5: Opinions on and attitudes towards genetic screening; A: Pre-natal screening (cystic fibrosis). Online, available HTTP: http://www.education.leeds.ac.uk/research/projects/young-peoples-understanding-of-and-attitudes-to-the-new-genetics [accessed 13 September 2013].

Lemke, J. L. (1990) *Talking Science: Language, Learning, and Values.* Norwood, NJ: Ablex.

Lemke, J. (2001) 'Articulating communities: sociocultural perspectives on science education'. *Journal of Research in Science Teaching*, 38, 3: 296–316.

Levinson, R. (2010) 'Science education and democratic participation: an uneasy congruence?' *Studies in Science Education*, 46, 1: 69–119.

Levinson, R. and Turner, S. (2001) *Valuable Lessons – Engaging with the Social Context of Science in Schools: Recommendations and Summary of Research Findings*. London: Wellcome Trust.

Lippmann, W. (1993)[1927] *The Phantom Public*. New Brunswick and New Jersey: Transaction Publishers.

Lippmann, W. (2009)[1921] *Public Opinion*. London: Lightning Source.

Lucas, J. R. (1979) 'Wilberforce and Huxley: a legendary encounter'. *The Historical Journal*, 22, 2: 313–30.

Macer, D. (2000) 'Where do we learn bioethics and avenues for bioethics education?' Proceedings of the UNESCO International Bioethics Committee Sixth Session, Volume II. Paris: UNESCO: 25–332.

Machiavelli, N. (2005) [1532] *The Prince*, trans. P. Bondanella, Oxford, UK: Oxford University Press.

Maienschein, J. (2003) *Whose View of Life?: Embryos, Cloning and Stem Cells*. Cambridge, MA: Harvard University Press.

Mansell, T. (1984) 'Some aspects of the history of biological education over the last two hundred years'. *Studies in Higher Education*, 9, 2: 97–111.

Martin, D. and Caldwell, S. (2011) '150 human animal hybrids grown in UK labs: Embryos have been produced secretively for the past three years'. *Mail Online*, 22 July. Online, available HTTP: http://www.dailymail.co.uk/sciencetech/article-2017818/Embryos-involving-genes-animals-mixed-humans-produced-secretively-past-years.html [accessed 8 November 2013].

Martínez-Gracia, M. V., Gil-Quílez, M. J. and Osada, J. (2006) 'Analysis of molecular genetics content in Spanish secondary school textbooks'. *Journal of Biological Education*, 40, 2: 53–60.

Marx, L. (2000)[1964] *The Machine in the Garden: Technology and the Pastoral Ideal in America*. New York: Oxford University Press.

Matterson, C. and Holman, J. (2012) *Informal Science Learning Review: Reflections from the Wellcome Trust*. London: Wellcome Trust.

Matzke, N. (2006) 'Design on trial: how NCSE helped win the Kitzmiller case'. *Reports of the NCSE*, 26 (1–2): 37–44.

Mauss, M. (2002) [1967] *The Gift: the Form and Reason for Exchange in Archaic Societies*, trans W.D Halls. London: Routledge.

McCombs, M. E. and Shaw, D. L. (1972) 'The agenda-setting function of mass media'. *Public Opinion Quarterly*, 36: 176–87.

McDonnell, O. and Allison, J. (2006) 'From biopolitics to bioethics: church, state, medicine and assisted reproductive technology in Ireland'. *Sociology of Health and Illness*, 28, 6: 817–37.

McKie, R. (2007) 'Flaming heck, first Brits were redheads'. *Sunday Times*, News Review, 7 January: 8.

McRobbie, A. (1994) *Postmodernism and Popular Culture*. London and New York: Routledge.

McSharry, M. (2009) *Schooled Bodies?: Negotiating Adolescent Validation Through Press, Peers and Parents*, London: Trentham Press.

Meillassoux, Q. (2008) *After Finitude: An Essay on the Necessity of Contingency*. London: Continuum.

Merton, R. K. (1973) *The Sociology of Science: Theoretical and Empirical Investigations*. Chicago, IL: University of Chicago Press.

Michael, M. (1996) *Constructing Identities: the Social, the Nonhuman, and Change.* London: Sage.

Michael, M., Grinyer, A. and. Turner, J. (1997) 'Teaching biotechnology: identity in the context of ignorance and knowledgeability'. *Public Understand of Science,* 6: 1–17.

Midgley, M. (1992) *Science as Salvation: Modern Myth and its Making.* London and New York: Routledge.

Midgley, M. (2001) 'Why memes?' in H. Rose and S. Rose (eds) *Alas Poor Darwin: Arguments against Evolutionary Psychology.* London: Vintage Books, 67–84.

Mol, A. M. (2002) *The Body Multiple: Ontology in Medical Practice.* Durham, NC: Duke University Press.

Mooney, C. (2005) *The Republican War on Science.* New York: Basic Books.

Moore, W. J. (1994) *A Life of Erwin Schrödinger.* Cambridge, MA: University of Cambridge Press.

Murphy, P. (2005) 'Young people's perspectives on genetics, identity and society using film and discussion', in H. E. Fischer (ed.) *Developing Standards in Research on Science Education.* London: Taylor and Francis Group, 165–70.

Murphy, P. (2008) 'Come on, be serious: positioning and framing in the power play of classroom-based reproductive and genetic technology debates', in A. R. Bell, S. R. Davies and F. Mellor (eds) *Science And Its Publics.* Newcastle: Cambridge Scholars Publishing, 57–77.

Murphy, P. (2010) *Nanotechnology: Public Engagement with Health, Environment and Social Issues: STRIVE Programme Report No. 61.* Wexford, Ireland: Environmental Protection Agency.

Myers, G. (2004) *Matters of Opinion: Talking About Public Issues.* Cambridge, MA: Cambridge University Press.

Nelkin, D. (1996) 'The Science Wars: responses to a marriage failed'. *Social Text,* 46/47, 14, 1–2: 93–100.

Nelkin, D. and Anker, S. (2004) *The Molecular Gaze: Art in the Genetic Age.* New York: Cold Spring Harbor Laboratory Press.

Nelkin, D. and Lindee, M. S. (2004) *The DNA Mystique: the Gene as Cultural Icon* (2nd ed.). New York: Will Freeman and Company.

Nisbet, M. C. and Mooney, C. (2007) 'Framing science'. *Science,* 6, 316: 56. Online, available HTTP: http://www.sciencemag.org.remote.library.dcu.ie/cgi/content/full/316/5821/56. [accessed 27 April 2013].

Noelle-Neuman, E. (1993) *The Spiral of Silence: Public Opinion – Our Social Skin* (2nd ed.). Chicago, IL: University of Chicago Press.

Nowotny, H., Scott, P. and Gibbons, M. (2001) *Re-thinking Science: Knowledge and the Public in an Age of Uncertainty.* Cambridge, UK: Polity Press.

O'Mahony, P. and Delanty, G. (1998) *Rethinking Irish History: Nationalism, Identity, and Ideology.* London: Macmillan Press.

O'Mahony, P. and Schäfer, M. S. (2005) 'The "Book of Life" in the press: comparing German and Irish media discourse on human genome research'. *Social Studies of Science,* 35, 1: 99–130.

Osborne, J. (2000). 'Science for citizenship', in J. Osborne and M. Monk (eds) *Good Practice in Science Teaching: What Research has to Say.* Buckingham: Open University Press.

Osborne, J., Duschl, R. and Fairbrother, R. (2002) *Breaking the Mould?: Teaching Science for Public Understanding.* London: Nuffield Foundation.

Osborne, J., Erduran, S. and Simon, S. (2004) 'Enhancing the quality of argumentation in school science'. *Journal of Research in Science Teaching*, 41, 10: 994–1020.

Ovid (1986) *Metamorphoses.* Oxford: Oxford Classics.

Parker, M. (1996) 'Biotechnology, children and ideal democracy', in European Commission, Cultural and social attitudes to biotechnology: analysis of the arguments, with special reference to the views of young people ('The BIOCULT Project'). Luxembourg: European Commission, 157–65.

Phelan, P., Davidson, A. and Cao, H. (1991) 'Students' multiple worlds: negotiating the boundaries of family, peer, and school cultures'. *Anthropology and Education Quarterly*, 22, 3: 224–50.

Pickering, A. (1993) 'The mangle of practice: agency and emergence in the sociology of science'. *American Journal of Sociology*, 99, 3: 559–89.

Pierce, C. (2013) *Education in the Age of Biocapitalism: Optimizing Educational Life for a Flat World.* New York: Palgrave Macmillan.

PISA (2012) Education Research Centre. Online, available HTTP: http://www.erc.ie/?p=55 [accessed 13 September 2013].

Pitkin, H. (1967) *The Concept of Representation.* Berkeley and LA: University of California Press.

Planned Parenthood of Souteastern Pennsylvania v Casey, Governor of Pennsylvana et al (1992) Online. 505 US Supreme Court 833, available HTTP: http://www.law.cornell.edu/wex/planned_parenthood_of_southeastern_pennsylvania_v._casey_1992 [accessed 10 September 2013].

Plato (1987) [360-380 BC] *The Republic.* London: Penguin Books.

Polkinghorne, J. C. (2004) 'The person, the soul, and genetic engineering'. *Journal of Medical Ethics*, 30, 6: 593–97.

Prime Time (2005) Commission reports findings on assisted reproduction, RTE1, 10 May.

Rabinow, P. (1996) 'Artificiality and enlightenment: from sociobiology to biosociality', in *Essays on the Anthropology of Reason.* Princeton: Princeton University Press.

Rabinow, P. and Rose, N. (2006) 'Biopower today'. *BioSocieties*, 1: 195–217.

Ratcliffe, M. and Grace, M. (2003) *Science Education for Citizenship.* Buckingham: Open University Press.

Ravetz, J. (2005) 'The post-normal science of safety', in M. Leach, I. Scoones and B. Wynne (eds) *Science and Citizens: Globalization and the Challenge of Engagement.* London and New York: Zed Books, 43–53.

Ravetz, J. (2006)*The No-Nonsense Guide to Science.* Oxford: New Internationalist Publishers Ltd.

Rawls, J. (1993) *Political Liberalism.* New York: Columbia University Press.

Reckwitz, A. (2002) 'Toward a theory of social practices: a development in culturalist theorising'. *European Journal of Social Theory*, 5, 2: 245–65.

Reid, G. (2012) 'The television drama-documentary (dramadoc) as a form of science communication'. *Public Understanding of Science*, 21: 984–1001.

Reiss, M. J. and Caccavale, E. (2008) 'Miracles, monsters and disturbances', in R. Levinson, H. Nicholson and S. Parry (eds) *Creative Encounters: New Conversations in Science, Education and the Arts.* London: Wellcome Trust, 48–63. Online, available HTTP: http://www.wellcome.ac.uk/About-us/Publications/Books/WTD028351.htm [accessed July 8 2013].

Reiss, M. J. and Straughan, R. (1996) *Improving Nature? The Science and Ethics of Genetic Engineering.* Cambridge, UK: Cambridge University Press.

Rodriquez, M. E. (2012) 'Eugenics' extension in the Spanish health care system through the prenatal diagnosis'. *CuadBioét*, 23, 77: 53–70. Online, available HTTP: http://www.ncbi.nlm.nih.gov/pubmed/22548657 [accessed 28 August 2013].

Roe v. Wade (1973) Online. (No. 70-18) 314 US Supreme Court, F.Supp. 1217. Available HTTP: http://www.law.cornell.edu/supct/html/historics/USSC_CR_0410_0113_ZS.html [accessed 10 September 2013].

Rose, C. (2003) 'How to teach biology using the movie science of cloning people, resurrecting the dead, and combining flies and humans'. *Public Understanding of Science*, 12, 3: 289–96.

Rose, N. (2007) *The Politics of Life Itself: Biomedicine, Power and Subjectivity in the Twenty-first Century.* New York: Princeton University Press.

Rose, S. (1997) *Lifelines: Biology, Freedom, Determinism.* London: Penguin Books.

Rose, H. and Rose, S. (2001) 'Introduction', in H. Rose and S. Rose (eds) *Alas Poor Darwin: Arguments Against Evolutionary Psychology.* London: Vintage, 1–13.

Rousseau, J-J. (1998) [1762] *The Social Contract.* Kent, UK: Wordsworth Classics.

Ryan, G. W. and Bernard, H. R. (2003) 'Data management and analysis methods', in N. K. Denzin and Y. S. Lincoln (eds) *Collecting and Interpreting Qualitative Materials* (2nd ed.). Thousand Oaks, CA: Sage Publications, 259–309.

Sadler, T. D and Fowler, S. R. (2006) 'A threshold model of content knowledge transfer for socioscientific argumentation'. *Science Education*, 90, 6: 986–1004.

Schatzki, T. R. (2001) 'Introduction: practice theory', in T. R. Schatzki, K. Knorr-Cetina and E. von Savigny (eds) *The Practice Turn in Contemporary Theory.* Routledge: London and New York, 1–14.

Schatzki, T. R. (2002) *The Site of the Social: a Philosophical Account of the Constitution of Social Life and Change.* Pennsylvania State University Press.

Schatzki, T. and Natter, W. (1996) 'Sociocultural bodies, bodies socio-political', in T. Schatzki and W. Natter (eds) *The Social and Political Body.* New York and London: The Guildford Press, 1–25.

Sebald, W. G. (2002) *The Rings of Saturn*, trans. M. Hulse. London: Vintage Books.

Selden, S. (1991) 'Selective traditions and the science curriculum: eugenics and the biology textbook, 1914–1949'. *Science Education*, 75, 5: 493–512.

Shapin, S. and Schaffer, S. (1985) *Leviathan and the Air-Pump: Hobbes, Boyle and the Experimental Life.* New York: Princeton University Press.

Sheldon, S. and Wilkinson, S. (2004) 'Should selecting saviour siblings be banned?' *Journal of Medical Ethics*, 30: 533–37.

Shilling, C. (1993) *The Body and Social Theory*, London: Sage Publications.

Shilling, C. (2005) *The Body in Culture, Technology, and Society.* London: Sage Publications.

Simmonneaux, L. (2000) 'Influence of cultural and disciplinary identity on the way teachers in agricultural education in France relate to knowledge in biotechnology'. *New Genetics and Society*, 19, 1: 23–47.

Simonneaux, L. (2001) 'Role-play or debate to promote students' argumentation and justification on an issue in animal transgenesis'. *International Journal of Science Education*, 23, 9: 903–27.

Simon, S., Erduran, S. and Osborne, J. (2006) 'Learning to teach argumentation: research and development in the science classroom'. *International Journal of Science Education*, 28, 2–3: 235–60.

Skloot, R. (2011) *The Immortal Life of Henrietta Lacks.* London: Pan Books.

Sloterdijk, P. (1988) *Critique of Cynical Reason*, trans. Michael Eldred. Minneapolis, MN: University of Minnesota Press.

Sloterdijk, P. (2011) *Bubbles: Spheres Vol. 1: Microspherology*. Boston, MA: MIT Press.

Slowther, S. (2008) 'Selection of embryos'. *Clinical Ethics*, 3: 60–2.

Smith, M. K. (2001) 'What is non-formal learning?' Infed.org. Online, available HTTP: http://infed.org/mobi/what-is-non-formal-education/ [accessed 25 August 2013].

Snow, D. A. and Benford, R. D. (1988) 'Ideology, frame resonance, and participant mobilisation'. *International Social Movement Research*, 1: 197–219.

Solomon, J. (1994) 'Knowledge, values and the public choice of science knowledge', in J. Solomon and G. Aikenhead (eds) *STS Education: International Perspective on Reform*. New York: Teachers College Press, 99–110.

Solomon, J. (1999) 'Meta-scientific criticisms, curriculum innovation, and the propagation of scientific culture'. *Journal of Curriculum Studies*, 31, 1: 1–15.

Soto Laveaga, G. (2009) *Jungle Laboratories: Mexican Peasants, National Projects, and the Making of the Pill*. Durham, NC: Duke University Press

Spencer, H. (1864–1867) *The Principles of Biology (2 vols.)*. London: Williams and Norgate.

Splice (2009) [Film] Directed by Vincenzo Natali, Canada/France/USA: Gaumont.

Spunout.ie (2013) (Homepage). Online, available HTTP: spunout.ie [accessed 5 November 2013]

Stock, G. (2002) *Redesigning Humans: Choosing Our Children's Genes*. London: Profile Books.

Stocklmayer, S., Gore, M. M. and Bryant, C. (eds) (2002) *Science Communication in Theory and Practice*. Nordrecht, the Netherlands: Kluwer Academic Publishers.

Strathern, M. (1996) 'Enabling identity?: biology, choice and the new reproductive technologies', in S. Hall and P. du Gray (eds) *Questions of Cultural Identity*. London: Sage Publications, 37–52.

Sutton, V. R.,. Bree, A. F. and van Bokhoven, H. (2010) 'Ankyloblepharon-Ectodermal Defects-Cleft Lip/Palate Syndrome', [Online]. in R. A. Pagon, M. P. Adam, T. D Bird, C. R Dolan, C.T Fong, and K. Stephens (eds) *GeneReviews*™ Seattle, WA: National Centre for Biotechnology Information. Online, available HTTP: http://www.ncbi.nlm.nih.gov/books/NBK43797/ [accessed 30 July 2013].

TalkOrigins.net (2005) Kitzmiller v. Dover Area School District Decision of the Court (Concluded) Archive. Online, available HTTP: http://www.talkorigins.org/faqs/dover/kitzmiller_v_dover_decision3.html [accessed 13 September 2013].

Tannert, C. (2006) 'Thou shalt not clone'. *EMBO Reports*, 7: 238–40.

Tarde, G. (2000)[1899] *Social Laws: an Outline of Sociology*, trans. H.G. Warren. New York: Kitchener.

Taylor, C. (1989) *Sources of the Self: the Making of the Modern Identity*. Cambridge, MA: Harvard University Press.

Thacker, E. (2003) 'What is biomedia?' *Configurations*, 11, 1: 47–79.

Thacker, E. (2005) *The Global Genome: Biotechnology, Politics, and Culture*. Boston, MA: MIT Press.

Theatre of Debate (2013) (Homepage). Online, available HTTP: http://www.theatreofdebate.com/ [accessed 10 September 2013].

Thompson, D. F. (2010) 'Representing future generations: political presentism and democratic trusteeship'. *Critical Review of International Social and Political Philosophy*, 13, 1: 17–37.

Thrift, N. (2005) 'From born to made: technology, biology and space'. *Transactions of the Institute of British Geographers*, 30: 463–76.

Thrift, N. (2008) *Non-Representational Theory: Space | Politics | Affect*. London and New York: Routledge.

Titone, R. (1958) 'The development of Italian educational philosophy in the 20th century'. *International Review of Education*, 4, 1: 313–26.

Tobin, J. (2000) *Good Guys Don't Wear Hats: Children's Talk about the Media*. Stoke-on-Trent, UK, New York and London: Trentham Books/Teachers College Press.

Torgersen, H., Hampel, J., Von Bergmann-Winberg, M-L., Bridgement, E., Durant, J., Einsiedel, E., Fjaestad, B., Gaskell, G., Grabner, P., Hiber, P,. Jelsoe, E., Lassen, J., Marouda-Chatjoulis, A., Nielsen, T. H., Rusanen, T., Sakellaris, G., Siefert. F., Smink, C., Twardowski, T. and Kamara, M. C. (2002) 'Promise, problems and proxies: twenty-five years of debate and regulation in Europe', in M. W. Bauer and G. Gaskell (eds) *Biotechnology: The Making of a Global Controversy*. Cambridge, UK: Cambridge University Press, 21–94.

Toulmin, S. (1958) *The Uses of Argument*. New York: Cambridge University Press.

Townsend, N. (1999) *The Gift* teachers' resource. Online, available HTTP: http://www.ytouring.org.uk/gifteducationpack/introf.html [accessed 2 July 2007].

Trench, B. (2001) 'Science, culture and public affairs'. *The Republic*, 3: 53–63.

Trench, B. (2008) 'Towards an analytical framework of science communication models', in D. Cheng, M. Claessens, T. Gascoigne, J. Metcalfe, B. Schiele, and S. Shi, (eds) *Communicating Science in Social Contexts: New models, New Practices*. New York: Springer, 119–38.

Turner, B. S. (1996) *The Body and Society: Explorations in Social Theory*. London: Sage Publications.

Turner, V. (1974) *Dramas, Fields, and Metaphors: Symbolic Action in Human Society*. Ithaca, NY: Cornell University Press.

Turney, J. (1998) *Frankenstein's Footsteps: Science, Genetics and Popular Culture*. New Haven and London: Yale University Press.

UNESCO (1945) UNESCO Constitution. Online, available HTTP: http://portal.unesco.org/en/ev.php-URL_ID=15244&URL_DO=DO_TOPIC&URL_SECTION=201.html [accessed 13 September 2013].

UNESCO (1997) Universal Declaration on the Human Genome and Human Rights. Online, available HTTP: http://portal.unesco.org/shs/en/ev.php-URL_ID=2228&URL_DO=DO=_TOPIC&URL_SECTION=201.html. [accessed 13 Sepember 2013.]

UNESCO (2003) *Report of the IBC on Pre-implantation Genetic Diagnosis and Germ-line Intervention*. Division of the Ethics of Science and Technology. Online, available HTTP: http://portal.unesco.org/shs/en/files/2397/10554294261ReportfinalPGD_en.pdf/ReportfinalPGD_en.pdf [accessed 8 November 2013].

Upstream Color (2013) [Film] Directed by Shane Carruth, USA: erbp.

Utton, T. (2004) 'The IVF twins who have five parents; grandmother aged 44 gives birth to her infertile daughter's children'. *Daily Mail*, 13 September, 30.

van Langenhove, L. and Harre, R. (1999) 'Introducing positioning theory', in R. Harre and L. van Langenhove (eds) *Positioning Theory: Moral Contexts of Intentional Action*. Oxford/Massachusetts: Blackwell Publishers Ltd, 14–31.

Verhey, A. (2003) '"Playing God" and invoking a perspective', in R. Sherlock and J. D. Morrey (eds) *Ethical Issues in Biotechnology*. Maryland, US: Rowan and Littlefield, 71–87.

Vivanco, L., Martínez, A. and Jouve de la Barreda, N. (2010) 'Bioethics and biojuridical evaluation of preimplantation genetic diagnosis in Spain', trans. NCBI *Cuad Bioét.* 21, 72:213–30. Online, available HTTP: http://www.ncbi.nlm.nih.gov/pubmed/20886913 [accessed 28 August 2013].

von Glaserfeld, E. (2001) 'The radical constructivist view of science'. *Foundations of Science*, 6, 1–3: 31–43.

Walsh, B. (2007) *The Pedagogy of Protest: The Educational Thought and Work of Patrick H. Pearse*. Bern: Peter Lang.

Warnock, M. (1985) *A Question of Life: the Warnock Report on Human Fertilisation and Embryology*. Oxford: Basil Blackwell Ltd.

Wellcome Trust (2000) *A Qualitative Evaluation of The Gift*. London: Wellcome Trust.

Whitehead, A. N. (1920) *The Concept of Nature: The Tarner Lectures*. Cambridge, UK, Gutenberg Project. Online, available HTTP: http://www.gutenberg.org/ebooks/18835 [accessed 13 September 2013].

Whitelegg, M. (1998) 'Literature review', in European Commission, *Cultural and Social Attitudes to Biotechnology: Analysis of the Arguments, with Special Reference to the Views of Young People* ('The BIOCULT Project'). European Commission: Luxembourg: European Commission, 59–81.

Whiteside, K. (2011) 'A Representative Politics of Nature?: Pursuing Bruno Latour's "Collective"'. Paper presented to the Western Political Science Association Annual Meeting, 22 April.

Whyte, N. (1999) *Science, Colonialism and Ireland*. Cork: Cork University Press.

Widdows, H. (2009) 'Border disputes across bodies: exploitation in trafficking for prostitution and egg sale for Stem Cell Research'. *International Journal of Feminist Approaches to Bioethics*, 2: 1, 5–24.

Willis, P. E. (1981) *Learning to Labour: How Working Class Kids Get Working Class Jobs*. Hampshire, UK: Gower.

Wilsdon, J. and Willis, R. (2004) *See-through Science: Why Public Engagement Needs to Move Upstream*. London: Demos.

Wolbring, G. (2008) 'The Politics of Ableism'. *Development*, 51: 252–58.

Wynne, B. (1992) 'Misunderstood misunderstandings: social identities and public uptake of science'. *Public Understanding of Science*, 1 (3): 281–304.

Wynne, B. (2005) 'Risk as globalising "democratic" discourse?: framing subjects as citizens', in M. Leach, I. Scoones and B. Wynne (eds) *Science and Citizens: Globalization and the Challenge of Engagement*. London and New York: Zed Books, 66–82.

YouTube (2000) Human Genome Announcement at the White House. Online, available HTTP: http://www.youtube.com/watch?v=slRyGLmt3qc [accessed 29 August 2013].

Ziman. J. (1996) 'Is science losing its objectivity?' *Nature*, 382: 751–54.

Zimmerman, C., Bisanz, G. L., Bisanz, J., Klein, J. S and Klein, P. (2001) 'Science at the supermarket: a comparison of what appears in the popular press, experts' advice to readers, and what students want to know'. *Public Understand of Science*, 10, 1: 37–58.

Index

For Product Safety Concerns and Information please contact our EU representative GPSR@taylorandfrancis.com Taylor & Francis Verlag GmbH, Kaufingerstraße 24, 80331 München, Germany